T0213447

Repair

Péter Érdi · Zsuzsa Szvetelszky

Repair

When and How to Improve Broken Objects,
Ourselves, and Our Society

 Springer

Péter Érdi
Center for Complex Systems Studies
Kalamazoo College
Kalamazoo, MI, USA

Wigner Research Centre for Physics
Budapest, Hungary

Zsuzsa Szvetelszky
Research Center for Educational
and Network Studies
Centre for Social Sciences
Budapest, Hungary

ISBN 978-3-030-98907-1   ISBN 978-3-030-98908-8   (eBook)
https://doi.org/10.1007/978-3-030-98908-8

This Springer imprint is published by the registered company Springer Nature Switzerland AG
The registered company address is: Gewerbestrasse 11, 6330 Cham, Switzerland

To our former friends, who did not know about the importance of repair.

Foreword

A number of years ago, I was on the faculty at a university where a job candidate gave a talk on his subject, housing policy. His talk focused on aging housing stock in ethnic neighborhoods in older industrial cities, and what policies ought to be used to address the issue. At the end of the session, one of my colleagues summed the talk succinctly: "Things get old. They wear out."

We are not used to thinking about cities as wearing out, but in a serious sense they can. Indeed, entire societies can get old and become obsolete—their economies, their governance structures, and their social systems. They need repair.

Érdi and Szvetelszky give us a broad framework for thinking about what they call "the Repair Society." Rather than throwaway our broken tools, or toys, or friendships, or even social systems, hoping to replace them with something better, we should consider the option of repair. Oftentimes repair is less expensive and less wasteful than throwing things away and can lead to more satisfaction and less regret.

Not all repairs are successful, and not all new replacements are improvements. In the case of urban housing wearing out, civic leaders replaced the vibrant ethnic neighborhoods with sterile housing projects and vast highways through a program called "urban renewal," generally making old industrial cities less pleasant and encouraging the race to the suburbs. Now as cities re-populate, and suburbs deteriorate, we can see the waste generated by throwing away cities and replacing them with suburbs, and then repeating the cycle.

Repairs, in Érdi and Szvetelszky's sense, must be adaptive to changing circumstances as the world evolves. Or even as local conditions evolve. When I was a young boy, our family moved to a small south Alabama town into a modest house. The house had no central heat, no air-conditioning, one small bathroom, and three small bedrooms. The kitchen was small, and the dining area was in the living room. The floor was linoleum glued to the concrete slab.

As the needs and income of the family increased, my parents could have moved into a more spacious house or "repaired" the existing one. They chose the repair option and remodeled the small one-car garage into a kitchen. Now the kitchen was large, but one could not hide that it had been a garage. And the car had to be parked on the street! So they built a cinderblock garage.

This repair was satisfactory for years. Then my parents added another addition, connecting the main house to the cinderblock garage. Their master bedroom had air-conditioning, but our rooms still did not.

While the house then had a Rube Goldberg look, it still stands today, with the same structure. Repaired, not replaced, and still useable.

Érdi and Szvetelszky teach us why the repair society is often a better option than a throwaway and replaces society. Not only do we keep what is good about the old, but we don't run the risk of replacing the old with something worse—as in the case of urban renewal. Sometimes replacement is necessary, but we ought to explore the repair option first.

The Challenge of Repair in Dynamic Systems

In the most innovative parts of the book, the authors extend the notion of repair to complex systems, such as political or social systems. Complex systems are harder to repair, because they are themselves changing and adapting in ways we don't always understand. But if we don't repair (or what we might call reform), the system can change in ways that can be hard to reverse, evolving toward a place where we don't want to be.

In laying out the issues in repairing of complex social systems, Érdi and Szvetelszky show us that it is possible for systems to repair themselves by reaching homeostasis, a system of stable dynamic balance. Minor repair can keep the system going for a long time. An intervention now and then can put the system back on track, like my boyhood home.

But it is also possible that we can wait too long to fix a dynamic system heading toward breakage. We may note that the system is operating in a maladaptive way, but ignore the problems because they don't seem too bad.

Such a system may be accumulating errors—that is, moving step by step away from the homeostatic state. These errors may themselves act by multiplying instead of adding, making the problem much worse very quickly. So timely repair, especially in complex systems, is necessary.

The more errors accumulate, the more resistance there can be to fixing it, and the more difficult it will be to repair it. In 1956, a young Black woman, Autherine Lucy, was admitted to the then-segregated University of Alabama. Resistance by white segregationists, including the use of mob violence, resulted in her expulsion, even though she had done nothing wrong. The editor of the local newspaper, Buford Boone, wrote, "Gradual change has been taking place. But sometimes change cannot be continued in slow steps. There must be a jump. When Autherine Lucy made the jump, all hell broke loose." Boone predicted that there would be a reaction leading to less justice for Blacks in the short run, which is what happened. Repair—in the sense of restoring social progress—was stymied for a time as a consequence of what he called friction. But he predicted success in the longer run, which also happened.

Those who understand the need to repair societies must be ready for the friction in the system, and even strong forces that would undo the progress that has been made because of the repair.

Érdi and Szvetelszky give us a framework for thinking about these issues and many more they discuss in this provocative book. They point to a general set of principles to apply for all systems, both simple and complex, both static and dynamic, from minor maintenance to major overhaul. They warn against the extremes: try to avoid the options of no repair and total replacement. I hope *Repair* will stimulate a broad discussion on the need for fixing things that are going wrong on an ongoing basis rather than waiting until a crisis makes the tool, or marriage, or political system not amenable to being fixed.

Bryan D. Jones
J. J. "Jake' Pickle Regents" Chair
in Congressional Studies
Department of Government
University of Texas at Austin
Austin, USA

Preface

This book, of course, is about repair. We, the authors, had several motivations for writing this book. Maybe the first was personal. We have seen and experienced too many broken relationships around us, as have many others. The question is with us: Should and could they be repaired?

The second one was very global. Many people now agree that something went wrong in the world. Food waste and hunger, cheap clothing for the rich manufactured in conditions close to slavery in another part of the world, climate crisis, and social inequality. More frequent natural and social disasters. (Surprisingly, our initial motivation was neither COVID-19 nor war. We started to chat about the book before the pandemic began and worked on its last part as foreign soldiers invaded Ukraine).

Third, while a generation lies between us, we both have vivid memories that Hungary was a poor country in our childhood, where "repair culture" was the social norm.

Fourth, we, individuals, belong to communities and social groups. Physically (workplace, gym, church) and virtually (political groups, old boy—and girl!— networks, formal and informal support groups, social media groups). At an annual party on December 30th, 2019, in the house of mutual friends in Pest (i.e., on the flat side of Budapest), Péter told Zsuzsa about his recent dream about a new book on repair. Zsuzsa became extremely enthusiastic and sent a dozen emails while Peter traveled back to the USA, mentioning aspects of social psychology and the evolutionary psychology of repair. We gave each other and ourselves a three-month deadline to determine whether we could write a co-authored book. Soon, we had consensus on the idea that people have very diverse resources. Objects (such as our cell phones and laptops), close and more distant individuals (like family members, colleagues, and even accidental acquaintances), communities, and global resources (including air and ocean waters). What will we do if the resources are impaired? The unintended consequence of economic and technological development was the emergence of a society centered around disposable products, with further implications related to environmental crisis. The central core of the book thus emphasizes that repair is a general resource management strategy.

The repair process is also about creating, developing our personalities and knowledge, and protecting the environment. Repair provides people with many positive opportunities to live happier lives.

We, the authors, integrated our disciplinary approaches, as our fields are complex systems theory and social psychology, respectively. It remains our little secret how we cooperated, and the Reader is only interested in the final result. In any case, we had numerous conversations, mostly by Zoom, and several in person. The book reflects our dialogue about the topic. In the overwhelming majority in the cases, we came to consensus. It happened that we decided not to synchronize how we refer to ourselves. Occasionally, we used the first-person to share our stories (accompanied by a parenthetical to indicate which author is speaking—e.g., I (P)), and sometimes we use the third-person to describe our opinions (e.g., we might attribute an idea to "Zsuzsa" in the third person). Please, accept it!

Repair converts scientific theories to conventional applications by raising and answering questions like: When should we attempt to repair something, and when it is better to save one's energy and let things go? What are the scope and limits of our ability to repair things? The "developed world" wastes a huge amount of matter and energy—should we remain this way, or should we throwaway the "throwaway" society? How can we repair broken communities after natural and/or social disasters? How should we handle planned obsolescence? Will the business model of the circular economy dramatically reduce waste? When may we expect that disturbances in our world can be tamed to replace the original state, and when should we accept the emergence of a "new normal?" Mostly, how can we avoid existential risk? Our hope is that we learn how to live a resilient life and how to design resilient technological and social systems at small, intermediate, and large scales. "The show must go on," as they say.

While the book takes examples from economics, physics, social psychology, cultural studies, ecology, and environmental studies and the theory of complex systems, the readership is not restricted to scientists, and it is intended as a general interest book. The book is certainly not written for people whose lives are in perfect harmony—they should not buy the book! The book has a multi-generational audience: "Zoomers," who are growing up in a world where it seems everything is falling apart; people in their 30s and 40s, who are thinking about how to live a fulfilling life; Boomers, who are thinking back on life and how to repair relationships. Still, the book is offered not to everybody but to those who are interested in connecting the natural and social worlds and to those who have good questions and want even better answers.

There are a number of excellent books discussing different aspects of repair techniques and strategies.

- *Repair: Redeeming the Promise of Abolition*, Katherine Franke (Haymarket Books, 2019). Franke's book is about repairing the damage done by history. She focuses specifically on black history in the USA, illustrated by two regions. The book covers a special topic—intergenerational, systemic racism and the white privilege at the heart of American society—and argues that reparations for slavery are necessary, overdue, and possible.

- *Fumbling Towards Repair: A Workbook for Community Accountability Facilitators*, Mariame Kaba and Shira Hassan (AK Press, 2019). Kaba and Hassan's workbook focuses on a special area of formal communication: interpersonal harm and violence. Through their work, they present possibilities for improvement through reflection questions, skills assessments, facilitation tips, helpful definitions, activities, and hard-learned lessons. Their works support experts who coordinate and facilitate formal community accountability processes.
- *Darning: Repair Make Mend*, Hikaru Noguchi (Hawthorn Press, 2020). Climate change is a complex issue: Smaller and larger steps can be taken to improve problems. The authors present the execution of one small step—darning—in an artistic way. In visible mending, darning is a personal opportunity to reduce consumption. The book goes through the issue in depth, which, while significant, is very specific.
- *To Repair the World: Paul Farmer Speaks to the Next Generation*, Paul Farmer (University of California Press, 2019). Social activist Paul Farmer shares a collection of charismatic short speeches that aim to inspire the next generation. The speeches focus on inequalities and problems arising from global challenges. The book is aimed at those who want to improve the current situation in the world from a humane point of view.
- *Repair Revolution: How Fixers Are Transforming Our Throwaway Culture*, John Wackman and Elizabeth Knight (New World Library, 2020). *Repair Revolution* chronicles the crisis of the throwaway society and the rise of repair cafés: nonprofit, volunteer organizations devoted to repairing electronics and household items for free. *Repair Revolution* explores the philosophy and wisdom of repairing and do-it-yourself, as well as the rising "Right to Repair" movement. It finishes with the practical details: online resources, fixes for common product malfunctions, and tips for founding a repair café.
- *Repairing Infrastructures. The Maintenance of Materiality and Power*, Christopher R. Henke and Benjamin Sims (MIT Press, 2020) explains the theory and practice of the role of infrastructure repair in our life. Similarly to our intention, they also emphasize the growing interest in repair and maintenance, but we adopt a broader perspective.

As it can be seen from the examples above and from a number of other books we have not listed here, countless objects, phenomena, and processes can be improved in a specialized way. However, a work that would set out the general principles for the universal application of repair in the light of the relevant knowledge of the natural and social sciences is not yet on the market. That is why we are convinced that our work fills this gap.

Péter is grateful to the community of Kalamazoo College, and specifically to his close colleagues, who have provided a friendly, intellectual atmosphere. He is also indebted to his colleagues at the Department of Computational Sciences at

Wigner Research Centre for Physics in Budapest. He also thanks the Henry R. Luce Foundation for letting him serve as a Henry R. Luce Professor.

During the COVID-19 times, Péter spent his ill-fated almost totally remote sabbatical at the Center for the Study of Complex Systems of University Michigan. Even though it was remote, he very much enjoyed interacting with colleagues there. Two undergraduate students, Joanne Lee and Drew Trygstad, spent an academic year to help with very extensive literature review. Together with graduate student Christian Kelley Péter had many fruitful conversations with them about what *Repair* should and should not cover. The students also made many comments on the manuscript. Two members of the Center, Patrick Grim and Efrén Cortés, gave very useful comments on the draft, too.

Zsuzsa is grateful to the Károli Gáspár University of the Reformed Church in Hungary, Budapest, where she received many useful and inspiring thoughts, suggestions, and ideas from her colleagues and students during the development of the manuscript.

We are grateful for comments, conversations, correspondence, and/or moral support from a number of colleagues and friends: Ágnes Cselik, Bálint File, Orsolya Frank, Vilmos Friedrich, Máté Gulyás, Gergely Horváth, György Miklós Keseru, József Lázár, Beáta Oborny, Hillary Rettig, Caroline Skalla (thanks also for helping with the figures), Marcell Stippinger, Róbert Tardos, Jan Tobochnik, János Tóth, Balázs Ujfalussy, and András Vajda.

Natalie Thompson, a former student of Peter, now a graduate student in the political science program at Yale University has been serving as our copy editor. She not only carefully edited the original Hunglish version but also gave comments on the drafts of each chapter from the bird's eye perspective and helped in designing the structure of the book. A big thank you, Natalie!

We are thankful to our Editor Thomas Ditzinger for his encouragement and always positive attitude.

While we worked on the book, we received a lot of support from our families, and we are grateful to them as well.

Kalamazoo, USA   Péter Érdi
Budapest, Hungary   Zsuzsa Szvetelszky
April 2022

Contents

1 Introduction 1
  1.1 Our Early Encounters in Repairing 1
    1.1.1 Bridges over the Danube 1
    1.1.2 Palaces of Miracles 2
    1.1.3 A Lenci Doll 3
    1.1.4 The Table that Stood the Test of Fire 3
    1.1.5 Aunt Ella, the Needlewoman 3
    1.1.6 Oh, My Sister! 5
    1.1.7 The Book in Perspective 5
  1.2 The Rise and Fall of the "Throw-Away Society" 7
    1.2.1 From Poverty to Abundance 7
    1.2.2 Planned Obsolescence 10
    1.2.3 Food Waste Versus Hunger 13
    1.2.4 Fast Fashion: Under the Spell of Cheap Clothing 16
    1.2.5 Consumerism 19
    1.2.6 The Throw-Away Society: Emergence and Call
         for the End 21
  1.3 Social Relationships as Resources 23
    1.3.1 Friends Versus Acquaintances 23
    1.3.2 Social Groups as Resources 25
  1.4 Lessons Learned and Looking Forward 30
  References 31

2 A Golden Age that Never Was 35
  2.1 The Myth of the Golden Age: Looking into the Past 35
  2.2 Historical Examples: The Golden Age Again and Again 37
    2.2.1 Greece 37
    2.2.2 Rome 39
    2.2.3 Judeo-Christianity 40
    2.2.4 China 41
    2.2.5 Japan 42

2.2.6   Greece Versus China: Some Comparisons   42
2.3   Golden, but to Whom? the Dutch Controversy   43
2.4   From the Golden Age to the Climate Crisis   43
2.5   The Golden Age of Babies   45
2.6   Messages from the Past   46
2.7   Lessons Learned and Looking Forward   47
References   48

3   Why Do Things Go Wrong?   51
3.1   The Meandering Pathways of Irreversibility   51
3.2   Wear and Tear   54
3.2.1   Spontaneous Glass Breakage   54
3.2.2   Overwork and Burn Out   55
3.2.3   The Road to Societal Collapse   57
3.3   Extreme Events and Predictability   59
3.3.1   Extreme Events   59
3.3.2   Too Much Growth Is Just Too Much   60
3.4   Patterns of Damaged Relationships   65
3.4.1   Maintaining Stable Relationships   66
3.4.2   Why Do Relationships Break Down?   67
3.4.3   Warning Signals   68
3.4.4   Destruction of Groups: Dissolution and Polarization   71
3.5   Lessons Learned and Looking Forward   72
References   73

4   The Pathways Back to "Normal"   75
4.1   Stability, Homeostasis, and Resilience   75
4.1.1   Stability   75
4.1.2   Feedback Loops Everywhere   78
4.1.3   Resilience   80
4.2   Back to Normal: Some Case Studies   86
4.2.1   Recovery from Burn Out   86
4.2.2   The Art of Restoring Buildings   86
4.2.3   Resilience After Hurricane Katrina: Where Are We
Now?   89
4.2.4   COVID-19: Rapidly Changing Perspectives   91
4.3   Right to Repair, Fight to Repair   92
4.4   Lessons Learned and Looking Forward   98
References   99

5   The Pathways Toward the New Normal   103
5.1   When to Attempt Repair and When to Let Go   103
5.1.1   About the Rise and Fall of Catastrophe Theory   104
5.1.2   Self-organization: Transition to the New Normal   105
5.1.3   Big Cultural Changes: They Started Yesterday,
and Maybe It Is Not Too Late   114

5.1.4 Release or Repair: Guidelines for Decision Making 116
5.2 Transition to the New Normal: Some Case Studies 119
5.3 Lessons Learned and Looking Forward 121
References 122

6 Repair the World! 125
6.1 From Ancient to Modern Perspectives 125
6.1.1 Mythical Origins 125
6.1.2 How to Avoid Existential Risk? 127
6.2 From Times Past to Modern Theory to Action 130
6.3 A Capsule History of Reforms 136
6.4 Toward a Repair Society 140
6.4.1 How to Repair our Resource Management Strategies 140
6.4.2 Think and Act Glocally! 143
6.5 Lessons Learned and Looking Forward 147
References 148

7 Epilogue: Toward a Repair Society 153
References 157

Index 159

Chapter 1
Introduction

Abstract The chapter starts with personal stories of the authors about their early encounters with repairing objects. Despite increasing inequality, the developed world now lives with the blessing and curse of the throw-away society. The "throw-away" society gradually emerged during the 1950 s as because disposable goods appeared. Middle-class people of the Western world had the economic freedom to replace them when necessary, even when it was not. We use a small fraction of the clothes stored in our closets, and we realize that clothes are intentionally produced to have shorter life spans than perhaps previously. Even though many people still live in food insecurity and explicit hunger, food waste has become enormous. Many of us (still a minority) feel that having more stuff does not necessarily make us happier. We are now facing the replace or repair dichotomy. To make the transition to repair society, people need a new perspective on resource management. By resource, we mean not only the things and gadgets we own but also human resources, such as family members, friends, and the small and large communities to which we belong.

1.1 Our Early Encounters in Repairing

1.1.1 Bridges over the Danube

Budapest is often described as one of the most beautiful cities in the world. The flat "Pest" is separated from the hilly "Buda" by the river Danube (Duna in Hungarian), and the two are linked by (mostly) historical bridges. While all these bridges are unique and have totally different styles, many were blown up at the end of World War II by retreating German sappers. Some fell into the Danube. The majority of the bridges were rebuilt several years after the war, but the ruins of one art nouveau bridge remained in the water for another two decades, as a reminder of the destruction of the war. The Elisabeth bridge—named after the Habsburg queen Elisabeth (Sissi), beloved by the Hungarians—was the only one that could not be reconstructed after World War II, so a new suspension bridge was built as a replacement. We, the authors, grew up in this city, but our experiences are separated by a generation, and our memories of the city are different. Péter was already a high school senior when the new bridge opened, some time before Zsuzsa was born.

Zsuzsa moved with her parents from a small industrial city to Budapest at the age of seven. Anyone who lives in a big city can witness something being repaired, renovated, and restored everyday. Since the bridges over the Danube were hastily reconstructed after the war, they were the subjects of frequent renovation. It was always a topic of discussion with the other schoolmates—which bridge would be partially closed that day for renovations? If none of the bridges were closed, on the left banks of the Danube, you could see the emblematic, mostly neo-Gothic, and oversized Parliament, where part of the surface was always being maintained by a renovation team.

Following the war, in addition to the bridges, a huge number of buildings lay in ruins due to the effects of heavy Allied bombing, followed by a long and bloody siege of the city by the Soviet Red Army. We (Péter and his friends, just young boys at that time) played soccer each day in the shadows of the partially bombed factories and warehouses during the decades of post-war reconstruction [1].

1.1.2   Palaces of Miracles

Especially if you play soccer, and even if you don't, shoes will get torn and wear away from sauntering around the neighborhood after school. Uncle Matuszka had a cobbler shop close to the fluid border between Angyalföld, (the now-disappearing working class's "Land of Angels") and Újlipótváros ("New Leopold Town," inhabited by middle-class intellectuals of Jewish origin). He spent the whole, long day mending shoes in his street-facing workshop with the help of his deaf assistant. The room was dark—the only internal source of pale light was a single bulb hanging from the ceiling by a wire. When I (P) visited, I always smelled the scent of a mixture of leather, shoe-shine, glues, adhesives, dyes, and cat pee.

Boots were worn in the winter, sandals in the summer. Girls had Japanese leather slippers, and as I just learned from a dear girlfriend from my college years, the white upper part (made from matte leather) was artificially treated by a piece of chalk.

Uncle Matuszka repaired and restored everything. Broken heels and worn-out soles were repaired. Holes were patched. Blown seams were resewn. His workshop was full of tools, hammers, knives of different sizes, and a variety of nail and tack pullers. There were sewing threads and straight and curved hook needles. There were hardwood lasts—holding devices shaped like a human foot. Nails and screws lived in a metal container. Uncle Matuszka used wooden pins, "pegs," to attach leather soles. First, he made a hole with the awl, a sharp-pointed piece of steel, then placed a dowel before hammering the pin. As I correspond with my high-school and college classmates (these correspondences became very intense and frequent during the pandemic), we all agreed that cobblers' small and dark shops are preserved in our memories as palaces of miracles. We also agreed that it was one place where we learned that things should be repaired.

1.1.3 A Lenci Doll

Zsuzsa's mother was born at the beginning of World War II and was not even a toddler when she received a felt doll as a gift. The baby's body was made of stuffed canvas and her head of rubber, with huge, blinking eyes and a small mane cap. She was a prototype of the Lenci dolls famoulsy produced by Elena (Lenci) Scavini and her company. The golden age of the Lenci dolls was between the two world wars. Knowing European history, you would not be surprised to learn that the factory was bombed in 1944.

The doll endured for a long time, and when time began to leave its mark, Zsuzsa's aunt restored her: She sewed a new dress from checkered canvas and replaced the old one. This second outfit saw signs of wear more slowly because the now-teenage girl played with it less and ultimately put it aside over time. When Zsuzsa was born, she inherited the doll, and one of the female members of the family (perhaps her grandmother) refreshed the doll's outfit again. Time passed, and the doll received a new dress every 15 or 20 years. Zsuzsa's daughters did not stop this tradition. It was last re-sewn about 10 years ago, and while the rubber head has lost its color, no further wear is visible on it.

1.1.4 The Table that Stood the Test of Fire

In the middle of our (Z's) living room stands a 100-year-old, ornately carved, dark brown oak table. Showy, but not very comfortable, it is an old family heirloom, so it commands an important place. There is a 20-centimeter black spot in the center of the tabletop: It is darker than the furniture itself and looks as if the surface was damaged. But nobody, not even the youngest members of the family, wants it to be repaired "because it should remain like that!"

During World War II, one evening in the winter of 1944, half a dozen Russian soldiers occupied the home of our grandparents. They were angry and hungry, but mostly they were cold. One of them quickly removed a bundle of grandmother's storybooks from the shelf (my mom was a child at that time), threw them on the table, and set fire to the colorful drawings. The books were burning, but the table stood the test of fire. The family still uses this piece of furniture, and the great-grandchildren will not hear of restoring it. They say the patch is history itself and, were we to repair it, that history would be falsified.

1.1.5 Aunt Ella, the Needlewoman

I (P) was a teenager when it happened that my Mom needed help running our house-hold smoothly. Aunt Ella, a needlewoman and a somewhat distant, but emotionally involved, relative came by our apartment every two months to mend clothes, patch household textiles, and do other ordinary sewing for us. Among other things, she placed colored patches on the elbows of my overused sweater (the only one I had, in addition to a blue sweatshirt, which every boy in my class had).

Aunt Ella was diligent, creative, and, as I remember, very talkative. I do not believe she had a high school degree since between the two wars only a small portion of girls attended high school, which was called gymnasium (but had nothing to do with gyms!). Not more than 25% of high school students were female at that time. For those who could not afford to learn, becoming a skilled seamstress or needlewoman was an attractive alternative since they were able to earn a reasonable amount of money.

She brought some tools in a big brown bag. Needles, threads, and a variety of scissors, including pinking shears, were the most important items. The latter was particularly remarkable since its blades were saw-toothed instead of straight. Of course, I recognized the thimble—a hard, pitted cup intended to protect one's finger from pinpricks—which I also saw in our metal ditty box, where mom stored buttons, clips, ribbons, and small pieces of canvas. However, we did not have a sewing machine, so Aunt Ella took home those pieces that were labor-intensive.

I remember clearly one of her famous creations: A shirt of my dad's that she created. My dad was the chief engineer at a historical leather factory in Újpest, a separate city that became part of Budapest after the war. He had two classes of shirts, Class B was used in the tannery, and Class A was reserved for visiting the authorities in the Ministry of Light Industry (textile and leather factories belonged to this authority, while machine factories and forges were supervised by the Ministry of Heavy Industry), attending the theater, and other special occasions. I don't remember when a shirt from Class A lost its status and became Class B, but I do remember the technology Aunt Ella adopted for repairing the shirts. She cut an isosceles triangle the size of about 20 centimeters from the back of a white shirt to be mended and fashioned this less-used material into a new collar after she removed the worn, original collar. The removed part was replaced by some ugly brown patch. The result was a Swiss shirt, as my Dad ironically called this type.

I believed that this procedure was born in poor, postwar Central Europe, but I was wrong. I was surprised and excited to read the beautiful article of the British writer Geoff Dyer in the *New York Times Magazine* [2]:

> She mended my trousers when they got torn from playing football (soccer) in the schoolyard and turned the collars of my and my dad's shirts, removing and resewing them so that the old, frayed exteriors were hidden from view and replaced by the perfectly preserved interiors. This was consistent with the key economic fact of my parents–lives: It was never worth their while to pay anyone to do anything they could possibly do themselves.

1.1.6 Oh, My Sister!

The fraternal relationship between Zsuzsa and her sister, Anikó, was very intimate and deep, despite the five-year age difference between them. However, this relationship changed when Zsuzsa had her first child at the age of 21—from then on, it became increasingly difficult for Zsuzsa to find common ground with Anikó. They started to move away from each other, which is no wonder, as they both did different things: While Zsuzsa went to university, took care of a baby, and managed a household, Anikó lived the life of a teenager and then started working as a typist in an office.

The center of Budapest came to life at that time: Anikó had many more opportunities to have fun in the early 90s than Zsuzsa had had a few years earlier. Their daily rhythm also differed: Anikó got up much later and went to sleep later than Zsuzsa. They met less and less often. Their shared past, however, never dissolved as Zsuzsa always remembered the common childhood dollhouse constructions and board games.

At the age of 25, Zsuzsa realized: Their relationship was no longer the same! She realized what she had to do to get closer to Anikó again. She accepted that Anikó was interested in rock bands and fashionable clothes, and when they met, she began talking to Anikó about her own topics. Although she was not inspired by what she heard, she thought that now was a time when she, as an older sister, had to take responsibility for the relationship. Years passed, and Anikó had a baby girl at the age of 26. In caring for the baby, she immediately sought out Zsuzsa, who was happy to give advice and trusted that they would grow closer to each other. Their relationship did become closer again—it became very practical, but Zsuzsa did not mind. She thought their relationship could be further improved, and her belief was soon confirmed.

Zsuzsa may have been around 35 years old, Anikó around 30, when their relationship changed again. Although their everyday lives and problems were different, they could build on the old foundation: They understood each other again. Although they argued a lot and agreed on far fewer things than previously, time and will did their part. Their relationship became personal again. But with differences, deviations, and changes. The desire to understand and accept one another brought back the old atmosphere in which they both happily relaxed, even if Zsuzsa's favorite movie was *Avatar* and her sister's *Star Wars*, even if they cooked quite differently, listened to different music, and read different books.

1.1.7 The Book in Perspective

This book endorses a new way of thinking about repairing impaired objects, relationships, communities, and even the world. The Reader (well, and the writers) often has to decide when to attempt to repair things and when it is better to save one's energy and let them go. We use the term "thing" in a very general sense: It might be a cell phone, car, marriage, friend, workplace, country, climate, or even the narrow and broad world. Things form our physical and social resources, and we see repair as a strategy for managing our physical and human resources.

We can only repair things that previously functioned well. It helps us identify strategies for repair if we have some ideas about how and why things went wrong, how various disasters emerged. Throughout the book, we discuss not only the scope but also the limits of our ability to repair.

Why the *scope* and the *limits*? The question of when to repair and when to replace objects is always with us. Many of us are suffering from the problem of broken friendships, and we ask ourselves whether or not they could (and should) be mended. A stopped clock can and should be repaired, but a burned-out light bulb must be replaced. A bulb is replaceable. It would be ridiculous, however, to throw out your whole crystal chandelier with 16 lights (part of your family inheritance) if just one of these lights does not function. But sometimes big companies adopt business policies that do not provide access to spare parts, and a whole new gadget must be bought. There can be recourse: The "right to repair" movement, which we will discuss in a later chapter, began a legal fight to allow consumers to buy replacement parts at a reasonable price in order to repair their broken devices. The book analyzes the whys and hows of making the transition *from a throw-away society to a repair society*.

Our goal is two-fold: First, we provide an integrative framework for the theory and practice of repairing. The framework is integrative since we examine repair mechanisms that occur at very different levels of biological and social organization, from molecules to inanimate objects, ourselves, our relationships, and our whole society. Second, we provide the Reader with guidelines for dealing with impaired systems in the 21st century.

We live in a complex world where globalization offers its blessings and curses. The 2008 financial crisis was a significant indicator for many that globalized economic structures were dangerous for resiliency. In the spring of 2020, we entered a new phase of learning how to adapt when horrible circumstances (in this case, a pandemic) arise. More and more often, things go wrong, and we are faced with a decision about how to respond. Specifically, we often have three options: First, we may try to repair things to return to the original, normal state. Second, we may decide that the previous state of things cannot be restored and that it is better to let things go or replace them with new ones. Third, many of us feel, especially now, that it is not enough to return things to normal, and we should grasp the chance to make things better than they had been. How to make the best decisions—this is what this book is all about. *Repair* will unify the perspectives of complex systems research and social psychology, applying scientific theories to everyday experiences with entertaining personal stories. But first, do we need to repair things? Yes, we do.

1.2 The Rise and Fall of the "Throw-Away Society"

To understand the importance of the theory and practice of repair, we have to recognize where we are now and how we got here. We describe the transition from poverty to relative abundance, then the pathway towards the throw-away society. Planned obsolescence and the emergence of industries like fast food and fast fashion were predominant factors influencing individual choices that ultimately produced negative, unsustainable effects.

What did our ancestors need for survival? The same as we do. Food and clothes on the one hand, mates and friends on the other. We need *physical and social resources*, and the history of humankind is a long narrative about the acquisition and management of these resources.

Food and clothes have been and still are the fundamental tools of human survival. Our ancestors gathered seeds, grasses, and fruits and then spent hours each day chewing them. Then we learned how to control fire (well, not speaking about forest fires …). It is plausible to believe that it was an accident when the flesh of a beast killed in a forest fire was found to be more enjoyable, safer, and easier to chew and digest than the usual raw meat.

The Agricultural Revolution, which started with the cultivation of plants and domestication of animals, is somewhat paradoxical. Without agriculture, we could not have culture and science. However, the birth of agriculture can also be interpreted as the beginning of the *exploitation of nature*. The Agricultural Revolution contributed to a dramatic increase in population density, but then, as people lived closer to animals so as to facilitate farming and herding, new diseases, and also more interpersonal conflicts, emerged. In addition, the majority of people were very poor.

1.2.1 From Poverty to Abundance

"The history of poverty is almost the history of mankind" [3]. One feature of poverty is the extreme shortage of food—that is, famine. There are ancient Roman records that date to 441 BCE of famines in Western Europe due to the fall of the Western Roman Empire. There were recurring famines in England and the European continent. Mass starvation was frequent from the early Middle Ages to the beginning of the Industrial Revolution. Famines still affected Europe even during the twentieth century. However, there is a fundamental difference in the causes: In the vast majority of cases, natural disasters and subsequent population pressure on resources were the primary causes of pre-industrial famines, but modern Europe-wide famines were the products of human actions like dictatorship and war [4].

As it is well-known, Thomas Malthus (1766–1834), the English economist, suggested in his essay [5] that a linear increase in food production implies an exponential increase in population size (as opposed to contributing to a lower increase in population and a higher standard of living). It was a pessimistic perspective. The general view was that Malthus was wrong, as he underestimated the velocity of technological change. Cheap and ample energy (provided mostly by the propagation of steam engines) and cheap and plenty raw materials resulted in both enhanced agricultural and industrial productivity during the Industrial Revolution. Food production also grew exponentially, and the exponent of food growth was larger than that of the population growth. Malthus was labeled as a false prophet in an article in the *Economist* in 2008 [6].

The steam engine was one example of the conversion of one type of energy (thermal) into another (mechanical). James Watt's (1736–1819) steam engine used feedback principles to control engine speed through self-regulation. (We will return to study feedback loops in Sect. 3.3.1) This was a big event in the history of technological development. Watt designed a centrifugal flyball governor for regulating the speed of the rotary steam engine, and automatic control generated a constant speed. The steam engine revolutionized the textile industry and transportation. Robert Fulton (1765–1815), an American engineer, developed a steamboat in 1807, which traveled between New York and Albany on the Hudson River, and George Stephenson (1781–1848) designed steam-powered railways, which crossed between Stockton and Darlington in the northeast of England, in 1825. While the distances of the first travels were relatively short, soon the extent of steam-based transportation increased greatly.

So machines and new (coal-based) energy sources became the driving forces of the technological changes that connected society's new lifestyles and organizations. Factories were established, since the new machines were much too large to house in a worker's cottage. The factories needed people. There was a fear among small textile workshop owners that machine-based factories would replace people, who would lose their jobs. The transition to factories can be seen as an example of what we call *creative destruction*, a concept to be discussed in Sect. 5.1.2. (It seems to be a recurring fear, since many of the opponents of machine automation express similar fears.)

A community known as the Luddites organized actions to destroy factories. New machines and investments compensated for short-term by lowering prices and enabling the sale of new products. From a historical perspective, the Luddites were wrong, and their action was not creative destruction: "If the Luddite fallacy were true we would all be out of work because productivity has been increasing for two centuries" [7]. Soon, there was a transition from rural societies to urban communities. This transition is related to free markets and capitalism, and new *classes*, the bourgeois and proletariat, appeared.

Returning to technology and industrial development, as they are the tools of the transition to abundance, we cannot overestimate the significance of the Bessemer process for the manufacture of steel. The new method made it possible to produce much cheaper, more purified steel. Most importantly, this steel was then used for large construction. The "robber barons," as the men who oversaw industry giants are sometimes called, like them or not, contributed very much to the birth of modern America by building industries of steel and railroads and integrating such infrastructure with oil and electricity.

The previous two centuries were very contradictory. While there were recurring famines in the nineteenth and twentieth century centuries, and we had (at least) two terrible world wars, the years between 1800 and 2000 showed the most prolific growth and economic development in humankind's history. Much of this progress stems from the Enlightenment.

The gross world product (GWP) has grown with dramatic velocity. Data and mathematical models show that GWP evolved at rates faster than exponential—what we call super-exponential. At exponential growth rates, acceleration (the speed of the change in velocity) is constant, and GWP would be infinite over an infinite time horizon. When acceleration itself is growing, then GWP would tend toward infinity during finite time [8]. Compared to the lingering misery of the past, human society rapidly transitioned from poverty to abundance. (More discussion about growth processes will follow in Sect. 3.3.2).

We don't necessarily suggest that models capable of describing the past can be adopted without modification to predict the future. However, it is not unreasonable to imagine that new technologies like artificial intelligence will play a similarly innovative role and increase growth, as tools like fire, the wheel, the steam engine, and electricity did. It is too early to see the mechanism, but new economic models [9, 10] suggest that the job-reducing effects of automation are overcompensated for by the creation of new jobs that are not vulnerable to automation. However, optimistic and pessimistic predictions can coexist. In the fall of 2019, the *European Economic Review* published a virtual special issue with the title "The Economics of Malthusianism: Can Malthusism return?" In an era of declining population growth, when the proportion of dependents might surge dramatically due to an increase in life expectancy, some kinds of Malthusian concerns might return [11].

The adage is always with us: "When the rich get richer, the poor get …" You can continue the sentence with the words "poorer" or "richer," depending on your perspective. The English romantic poet Percy Bysshe Shelley (1792–1822) formulated the principle "The rich get richer and the poor get poorer." We believe that it might be true now for some, but not all, regions of the world. However beautiful is *New York Times* columnist Thomas Friedman's vision [12] of a world flattened by ubiquitous telecommunications, the world still is not flat! But it is certainly flatter than it was. It would be difficult to deny that the increase in inequality, mostly in the Western world, is very disturbing. However, as Steve Pinker, among others, argues, the tendency of global inequality is declining, since the growth rates of the poorest countries exceed those of the richest [13]. So, poorer people in poor countries might be seen as the winners of globalization. (Even though the payment in the garment industry, for example, is very low by Western standards, it is much larger than what workers made earlier as rural, generally small-scale farmers.) Of course the rich people in the richer countries are the biggest winners. The lower middle class of Western countries seem to be the losers, and they have become angry and susceptible to supporting populist politicians as a result. To formulate a balanced view, we should also emphasize that, more often than not, the development looks to be global, and it happens through the massacres, abuses, and exploitation of non-European countries and people in poverty. Pinker's statement is not factually wrong, but the celebrated rapid decline in poverty rates among impoverished countries really masks three trends (not necessarily mutually exclusive): (1) poverty increases (or shows little change) in some regions, (2) poverty declines in a slow-to-moderate (and yes, sometimes strong!) fashion in most regions, (3) and poverty strongly declines in the mammoth nations of India and China.

1.2.2 Planned Obsolescence

The concept of "planned obsolescence" [14] has been in the United States since at least the 1920s, when General Motors (GM) introduced a new production and marketing strategy. GM started to design and construct new models for just a single year, so last year's cars seemed to be out of date almost by the time they were released. The strategy worked and appeared in other sectors as well. If a product is old-fashioned, or does not work well after a certain period, the consumer seeks new items. Built-in, or planned, obsolescence can be implemented through two different mechanisms. First, a new, superior model appears and advertisements convince people to buy the new model. Second, the product is intentionally designed to be nonfunctional within a planned period of time. In both cases, consumers prefer the next generation of products to the older ones.

The main social reason that planned obsolescence worked was the rapid emergence of a strong middle class during the 1950s, which is sometimes called the "Decade of Prosperity." The purchasing power of the median American family grew by 30% during that time [15]. Progress in science and technology, combined with the availability of cheap oil from U.S. wells, were the driving forces of the increase in industrial productivity. Europe and Asia were still recovering from World War II, so America did not have peer competitors. People who lived in poverty during the Great Depression and World War II were very motivated to buy, and suddenly they were able to purchase, bigger cars, larger houses, and better and longer educations. In addition, disposable goods appeared, and people had the financial means to replace products.

From the Centennial Light to the Not-so-long-lived Bulbs

Commercially viable light bulbs were invented by Thomas Edison (1847–1931). In the first three decades after their invention, carbon filaments were used. The Centennial Light is the world's longest-lasting light bulb—it has been burning since June 1901 and is located in Livermore, California. It is a hand-blown, carbon-filament light bulb. While the bulb has been "off" several times, generally due to human intervention, it has never ceased to be functional. Since bulbs produced later had a much shorter life duration than that of the Centennial Light, they are frequently referred to as a paradigmatic example of planned obsolescence.

The Phoebus Cartel

The Phoebus cartel was formed in 1924 to manage and control the design and engineering of shorter-lived light bulbs. At that time, tungsten filaments were adopted with increasing frequency. An excellent paper describes the history of the *grand light bulb conspiracy* [16]. The goal of the bulb industrialists was a systematic reduction in the capacity of light bulbs from 1500–2000 hours of functionality to 1000. It is interesting to see that researchers had to find conditions that would *reliably* reduce the lifespan of the bulbs. Phoebus was a global cartel, including General Electric and Tokyo Electric, alongside the big European companies, such as Germany's Osram, the Netherlands' Philips, France's Compagnie des Lampes, Hungary's Tungsram, and Britain's Associated Electrical Industries. While the cartel managed to keep artificially elevated prices for a while, competitors did emerge to provide cheaper, often lower-quality, goods. Specifically, while Tokyo Electric was a member of the cartel, small, family-owned Japanese workshops produced nearly hand-made, cheap bulbs, mainly for the international market. The cheap bulbs were not necessarily financially favorable, since the low price was overcompensated for by increased consumption. As World War II started, the coordination necessary to keep the cartel alive was impossible, and the first global agreement to implement global planned obsolescence was nullified in 1940.

Planned Obsolescence is With Us, With No Intention of Leaving

Over the course of three decades, I (P) have had about 10 digital notebooks and laptops. In most cases, I decided to change devices when the cost of repair was not too far from that of buying a new, more technologically advanced model. I believe I still have at least four of the old devices, so technically they don't contribute to e-waste. Moore's law says that transistor count will double every 24 months, so generally the new laptop was of a higher technical level. I have been living in the world of Linux operating systems, now in the Ubuntu 20.04 version. While I don't believe the laptops were planned to have such limited reliability, the reality is that I am almost sure the Reader's experience is not very different from my own. (Two weeks after I wrote this paragraph, I returned to it to complain that the smartwatch I use to track and improve my fitness suddenly died after just 18 months of use.)

The surprising reality: Unrepairable laptops and smartphones are one symptom of the throw-away society in which we live. It is still somewhat surprising that the business model—based on batteries that users cannot replace—worked. We consumers could not resist buying these new gadgets, even though we were supposed to throw out the whole device when the battery failed. Not only do batteries die out, but we have also learned that sometimes things like operating systems or apps can suddenly no longer be upgraded.

A Balanced View

With some mixed feelings, we are inclined to agree with others' opinions [17] that the throw-away society has beneficial aspects as well.

First, the rapid turnover of goods correlates with the creation of jobs. *Second*, there has been a huge increase in the availability of relatively cheap goods for many people, not only in wealthy countries but also in developing countries. While we cannot deny that more people have had a better quality of life as a result of our consumer model than at any other time in history, it is also responsible for global warming and toxic waste.

Throw-away culture generates huge amounts of waste. As environmental consciousness continues to expand, consumer goods might become less disposable. Google initiated *Project Ara* to develop a modular smartphone device that lets you easily swap out components that are either broken or in need of an upgrade, but the company canceled the project a few years after it began.

As we throw away machines and devices, the result is a huge mountain of e-waste. It is worth to see the numbers published in the Global E-waste Monitor 2020 [18]:

"In 2019, the world generated a striking 53.6 [megatons (Mt)] of e-waste, an average of 7.3 kg per capita. The global generation of e-waste grew by 9.2 Mt since 2014 and is projected to grow to 74.7 Mt by 2030—almost doubling in only 16 years. The growing amount of e-waste is fueled by higher consumption rates of [electrical and electronic equipment], short life cycles, and fewer repair options. Asia generated the highest quantity of e-waste in 2019 at 24.9 Mt, followed by the Americas (13.1 Mt) and Europe (12 Mt), while Africa and Oceania generated 2.9 Mt and 0.7 Mt, respectively. Europe ranked first worldwide in terms of e-waste generation per capita, with 16.2 kg per capita. Oceania was second (16.1 kg per capita), followed by the Americas (13.3 kg per capita), while Asia and Africa generated just 5.6 and 2.5 kg per capita, respectively."

Value Engineering

"Value engineering" is a somewhat "politically correct" version of the concept of planned obsolescence. The goal of this design process is to preserve the function of a product while reducing the costs of production by using as little and as cheap material as possible. For those of us who grew up behind the iron curtain, the East German car the Trabant, arguably the worst automobile ever made, is the archetype of the concept. It utilized what was called a two-stroke engine, which was obsolete even at the time of its invention. It used to be said that two people were needed for its construction—one to cut and one to glue, as it was made from plastic, and many jokes were made about its quality and value. Here is an example:

Q: How do you double the value of a Trabant? A: Fill up the gas tank.

We were not totally right. Value engineering became a formal design concept during World War II when there was a shortage of everything, from raw materials to human resources. To ensure the continuity of the production process, the engineers at General Electric had to find both different materials and simpler (cheaper) procedures. The goal was to maintain the level of functionality *without* reducing the quality. Function should be clearly defined: "generate light," "store images," or "transport passengers on a usually four-wheeled vehicle by using an internal-combustion engine using a volatile fuel." Engineers found much better solutions than what we experienced with our Trabants.

1.2.3   Food Waste Versus Hunger

A number of my (P) Facebook friends like to share idyllic pictures of food consumed in the company of our loved ones. I do too. Many of us have some superficial interest in the origin and means of processing the food we consume. We like to see "cage-free" labels or know that the products are local. Still, in the United States, about 10% of products come from the farmer's market, and 90% comes from outside of local communities [19]. About half of the consumable food in the United States is thrown away, and wasted food is the single largest component of American landfills. The flip side of the coin is that one in seven Americans experience hunger, and during the COVID-19 pandemic, that number became one on six (so more than 50 million Americans lacked food security in 2020). Brenda Ann Kenneally, a photojournalist documented the food insecure lives of millions in several moving and difficult-to-forget photos [20].

It is easy to waste, isn't it? It would be difficult to deny that a typical source of food waste is "overbuying." Not just because we forget about our reserves at home but also because we simply misjudge our needs—we get more than we need or have run out of. Or we just change our minds as picky people: What we liked in the store and put in our shopping basket no longer appeals to us at home, and we throw it out.

We waste in a variety of ways: We often buy food when there is still enough in the fridge because we just forgot about it. For example, we might not know that we have it because we store it irregularly. There are a lot of packages lurking on top of each other in our pantries, so an expiration date can easily escape attention. A lot of recently expired food in unopened packaging lands in the trash everyday. Everyone is afraid of salmonella and its immediate consequences, but the long-term consequences of food waste are more difficult to see. But irregular, less-than-thorough cleaning of the refrigerator can also ruin food prematurely.

Recently, Chinese President Xi Jinping launched a "Clean Plate" campaign against food waste [21]. Statistics suggest that at least 17 million tons of food went to waste in China annually in the last several years [22]. We have yet to see the results of the Chinese attempt, which aims to establish a social environment where waste is shameful. More precisely, waste is not only shameful, but the lack of appropriate recycling of food waste is also penalized by the newly introduced social credit rating system.

In India, 25% of the fresh water used to produce food is ultimately wasted, even as millions of people still lack access to drinking water. Big weddings and huge parties are big contributors to Indian food waste [23]. There is a chronic imbalance of food distribution.

Why, why, and why? If individually we feel guilty about behaving this way, why do we have huge quantities of food waste in very different parts of the world? First of all, whatever we say after paying in the supermarket, food is still cheap. Second, we don't buy "ugly" foods. We often only like to buy produce that has an aesthetic quality. Manicured food is identified with food safety. As we are able to make excellent pictures with our cell phones, the so-called "camera cuisine" influences what we order to eat. There is a new profession—food stylist—that prepares photogenic food for photography, video, or film. As shoppers, we don't buy imperfect-looking fruits and vegetables. On the supply side, crooked produce does not even make it to the shelves—it is thrown out before it makes it to the store [24]. This is unreasonable, since biologically, the flavor of the food is more important than appearance. But data show that waste and hunger are rising on parallel tracks. It is not painless to accept the coexistence of food waste and hunger.

Once again: We think fresh foods are beautiful, while less fresh—withered, browned— foods are ugly. "We eat with our eyes," as some have said. The color and fragrance (turbocharged with state-of-the-art technology during production), the texture, the shape, and the packaging of the product all affect our most important senses: our eyes. It is estimated that we get 80% of the information we process from the outside world with our eyes, so there is a stake in the competition to be beautiful. A lot of goods are finished works of art: Artists and scientists design every detail, as well as the overall effect. In countless cases, we throw the unsightly in the trash: "It may still be good, but I won't eat it anymore."

While it is easy to write that our culinary culture should be subject to rethinking and repair, some elementary steps have been made for a long time. In a sentiment probably dating back to antiquity, people feel some moral obligation to help hungry individuals eat. The soup kitchen, as an institution, emerged at the end of the eighteenth century. While the Industrial Revolution created an increase in overall prosperity, it also exacerbated inequality. The traditional life of many poor people was disrupted, and the number of hungry people increased. Sir Benjamin Thompson, also known as Count Rumford, is credited as an early champion of hunger relief and established the first modern soup kitchen. The concept became extensive in the United States during the Great Depression. The infamous gangster labeled "Public Enemy Number One," Al Capone, opened a soup kitchen that "served breakfast, lunch, and dinner to an average of 2200 Chicagoans every day" [25].

Modern food banks are non-profit organizations and are instrumental in helping those people who cannot afford to purchase sufficient food. The modern era of food banks started in 1967, when St. Mary's Food Bank Alliance was founded in Phoenix, Arizona, by John van Hengel, who later founded the organization Feeding America to popularize the concept of food banks throughout the whole country. At that time, both politicians and the media drew attention to the issue of hunger, which led to the introduction of the Special Supplemental Nutrition Program for Women, Infants, and Children (WIC) and Supplemental Nutrition Assistance Program (SNAP), generally referred to collectively as the food stamp program. Food insecurity soared during the pandemic, and as food banks rely on personal donations, these contributions shrink during difficult times.

There are now companies like Imperfect Foods and Misfit Market, which purchase and deliver cosmetically challenged produce to consumers. I (P) clicked on the website of Imperfect Foods after their ad appeared on my Facebook page, and soon we became enthusiastic Imperfectionists. We discussed the food waste problem over dinner in my house (Hungarian goulash was served and consumed) with our neighbor, a professor of sociology/anthropology. She worries that while Imperfect Food is an appealing initiative, it may create competition with food banks, which have previously been able to acquire imperfect foods and provide them to those who need it.

Traditional housewives have always been creative in secondary processing. The leftovers of potato soup (the Central European version is different than what we have in the United States or Japan) can be potato souffle, and tomato salad could be converted into bruschetta or gazpacho. We can transform and improve leftovers to make our diet varied, and most importantly, so as not to throw away so much. It is also recycling, although not industrial, but individual and more creative. Last but not least, it is an innovative opportunity for ourselves.

The title of a tale by Danish writer Hans Christian Andersen is "The Girl Who Trod on the Loaf." In this story, the little woman throws the loaf she is supposed to give to her poor family on the ground so she can step on it and keep her slippers clean from the mud. Today, entire streets could be "paved" with bites thrown in the trash for various reasons. The issue is not only with bread, but also with cold cuts, cheeses, packaged sauces, and other delicacies. We don't not necessarily make rational decisions about buying and storing food, and we have not yet addressed the somewhat neglected activity of food recycling.

Throwing away food goes far beyond the specific loss of food or money. It involves wasted animal feed, pesticides, the cost of fuel and energy for production and transport, and the cost of storage, all of which add up to tangible losses for the consumer. Few are aware of this, as then we would have to face the hard-to-digest fact that we regularly harm ourselves. Over-consumption and overproduction can eventually become a trap for each other: Could it have become our subconscious need to throw out food?

A few years ago, there was still a common phrase in the recipe sections of magazines: Just eat and cook completely intact fruits and vegetables! Today, due to the impact of what might be called the bio-lifestyle, we read this less often, and the "ecological approach" overwrites the old advice. Now the typically middle class Reader of life-style magazines boldly cuts a little bump out of the otherwise flawless pieces instead of throwing the whole thing in the trash. One little strategic suggestion from Zsuzsa: Next time buy a smaller fridge! Less food will be ruined because you will have less space!

Environmentalists, business owners, and information technologists have all implemented new methods of fighting against food waste. Some restaurants, such as Brooklyn's *Rhodora*, are based on a moral principle of no-waste. Here are a number of applications worldwide aimed at reducing food waste [26]: *Karma* helps us to discover the cheaper offers of restaurants and bars near us at the end of the day. *Farmdrop* connects with local farmers so our footprint of buying and consuming will be smaller. With *Olio's* support, we can find partners nearby to whom we can hand over our leftovers. *Foodcloud* partners with supermarkets to help unsold food reach charity organizations and those in need. *Giki* provides information on the ethical and sustainable practices of distributors.

Michelin recently introduced a new distinction, the Green Star, to award restaurants that prove successful at reducing waste, preserving natural resources, and protecting endangered species. Thirteen Japanese restaurants have received this award in Kyoto, Tokyo, and Osaka [27]. (It is interesting, but not very surprising, that in Tokyo, the award has mostly gone to French restaurants, while in Kyoto, traditional Japanese ones have been granted the honor.)

We now have more and more opportunities to improve the process of waste reduction. Every little step could contribute to significant change, and every gram of food saved betters our own financial situation and sustainability. If we start paying attention to the shelves of the fridge, the expiration date on the packaging, and the contents of the pantry, we can experience a remarkable transformation in just a few weeks or maybe even in a few days. Our first feeling might be that we are sorry to throw away the cake that has not yet been eaten since we have to leave in the evening, so we can look for a neighbor to whom we can offer the leftovers. The second feeling is creativity: Yes, the apple is no longer as tight and red as it was when we bought it five days ago, but next to the meat, a steamed apple will be perfect.

As we worked on this chapter, we were happy to see that "The Norwegian Nobel Committee has decided to award the Nobel Peace Prize for 2020 to the World Food Programme (WFP) for its efforts to combat hunger, for its contribution to bettering conditions for peace in conflict-affected areas and for acting as a driving force in efforts to prevent the use of hunger as a weapon of war and conflict" [28].

We, as a society, are in the process of rethinking our entire food system—how we produce, distribute, and consume food. The stakes are our own health and that of the environment, the climate, and society, both in a physical and social sense.

1.2.4   Fast Fashion: Under the Spell of Cheap Clothing

Wearing clothing is purely human (even though some humans provide clothes for their pets). The textile industry benefited very much from the transition from inefficient small-scale production to mass, factory-led production that occurred during the Industrial Revolution. Edmund Cartwright was responsible for a major innovation—the power loom—at the end of the eighteenth century, which was the driving force of many of these changes.

People originally used natural fabrics, from silk to cotton to wool to linen, made from organic materials. Human creativity naturally asked the question: Can we prepare artificial fabrics with better qualities that can be made more cheaply? New materials, such as synthetic fibers, were invented in the early twentieth century. PVC was patented in 1913, and it became popular since it proved to be water-resistant and highly durable. In 1935, a new fiber was synthesized in the DuPont Chemicals laboratory: nylon! It reached the market a few years later and was used for military purposes—famously, for parachutes. Soon, it became very popular as a silk replacement. Nylon stockings were a huge commercial success. (Behind the iron curtain, no nylon stockings were produced. As a huge football fan, Péter cannot resist mentioning that the heroes of the Hungarian "Golden" soccer team were permitted to smuggle nylon stockings for resale after they won 6-3 against England in Wembley Stadium on a foggy November afternoon in 1953. We know that not every Reader is emotionally attached to this event.)

Natural versus Synthetic: Not Angel versus Devil

Diane von Fürstenberg, the iconic fashion designer, explains in her MasterClass that natural and synthetic fibers each have their advantages [29]. Natural fibers might be plant-based, like cotton, jute, and linen, or animal-based, like silk and wool. Generally, they are considered eco-friendly, water-absorbent, and durable. Synthetic fibers are the product of science and technological development. A chemical process, called polymerization, may lead to artificial or synthetic fibers. They are much cheaper than natural fibers and are more appropriate for mass production. Specifically, polyester is created from coal and petroleum. Since it does not absorb water well, it is not really recommended for rainy summer weather.

We don't believe that chemists and engineers, who discovered the technological process leading to synthetic fibers, should be blamed, even though the unintended consequences of their innovations are now well known. Eight million metric tons of plastic enter the oceans per year [30]. Synthetic fibers contribute to pollution, and the formerly neglected role of microplastics has recently gained recognition as an important element of ocean pollution. So, in addition to whole plastic products—like cups, shopping bags, and straws—that are thrown into the water, plastic particles less than 5 millimeters in length leave our washing machines day by day. Neither filters in washing machines nor in sewage treatment plants are able to catch them. Fish exposed to microplastics have lower reproduction rates [31]. It is also documented that the cycle is complete, and the consumption of fish leads to a microplastic transition from fish to humans.

Products with labels like "biodegradable" or "compostable" now look attractive to environmentally conscious consumers. However, the physical and chemical conditions in which these products degrade vary. For example, corn-based plastic is compostable only at high temperatures and when exposed to specific moisture levels. So it most likely will not actually degrade in a landfill. A recent prediction from a group of scientists from an impressive number of countries (Australia, Canada, Indonesia, the Netherlands, New Zealand, the United Kingdom, and the United States) found that "substantial reductions in plastic-waste generation can be made in the coming decades with immediate, concerted, and vigorous action, but even in the best-case scenario, huge quantities of plastic will still accumulate in the environment" [32–34].

In the twenty-first century, there has been a dramatic decrease in the lifespan of clothing. As a consequence of overseas production, clothes have become much cheaper (yes, it sounds like something good), and we have three times more clothes than people did in the 1960s. We may call it over-consumption. "Fast fashion" not only contributes to environmental but also human catastrophes. For example, a man-made disaster killed more than 1000 workers in Dhaka, Bangladesh, when a garment factory collapsed in 2013.

Why? The answer is not surprising: The factory was built with substandard materials. The world's top brands and retailers benefit greatly from the 5000 garment factories in Bangladesh, where workers have the lowest wages in the world of garment production [35]. The average worker's earnings are approximately 50$ a month— roughly the price of one of the pairs of pants they assemble for sale in the United States or Europe. A report by Open Society, written five years after the terrible accident, summarizes the responses to the incident [36]. The Accord on Fire and Building Safety in Bangladesh was created and signed by over 200 apparel brands and retailers to support building a safe and healthy garment industry. While the agreement was originally approved for five years and has been extended, there are serious concerns that, as our memory of the horrible accident fades, the commitments of the corporations to improve working conditions will also disappear.

The book *Fashianopolis* by Dana Thomas documents the unsustainability of the textile industry's current practices and discusses options regarding new technologies and business models [37]. Are we institutionally powerless? Slow-fashion, the return to small-batch production, and the transition to technologies that allow waste materials to be recycled seem to be tools of change. Are we individually powerless? We have just a piece of simple advice: Buy fewer clothes and wash only when really necessary!

Kate Wood, a beauty writer, summarized beliefs we share in a list titled "8 Reasons to Rethink Fast Fashion" [38]:

1. Fast fashion exploits overseas workers;
2. Fast fashion contributes to the decline of U.S. manufacturing;
3. Fast fashion also exploits U.S. workers;
4. Fast fashion is environmentally disastrous;
5. Fast fashion can wind up costing you more than "real" clothes;
6. Fast fashion's low quality changes how you think about clothes;
7. Fast fashion collaborations trick you into paying for the name; and
8. Fast fashion distorts your sense of value.

Renting clothes, as opposed to owning them, seems to be another viable strategy and an example of the new sharing economy. (Of course, it has its limits: buying clothes secondhand is okay but renting jeans and t-shirts is not realistic.) The model of the sharing economy is defined as a peer-to-peer system of acquiring, providing, and sharing access to goods and services. Among a few others, the company Rent The Runway permits its members to rent outfits, mostly for special occasions. Some other companies are piloting clothing rental subscriptions as well: Banana Republic has a program, as does Ralph Lauren. Clothing rental companies suffered during the pandemic, but we may expect them to bounce back as people return to offices, restaurants, and other venues, and they may even become more popular.

What we see these days is that there is a slowly, but surely, changing attitude in society. Big companies have made some high-profile efforts to be seen as eco-friendly. H&M recently announced an in-store recycling system installed in Stockholm, which they advertise as a "recycling revolution" [39].

Newer data suggest that the secondhand clothing market is booming and might help to reduce the sustainability crisis in the fashion industry [40]. Thrift stores and resale platforms enable consumers to buy and resell secondhand clothes, and they are touted as possible tools to reduce the dominance of fast fashion. Young Reader, your own consumer behavior will decide whether or not the digital resale market will be the "Next Big Thing" in the fashion industry.

Taelor is a California-based fashion industry start-up founded by Anya Cheng. The goal is to help busy men look good without the obligation of buying clothes. Taelor summarizes the problems facing Millennials living in the "wealthy" part of the world as relating to uncertainty about how to dress, lack of time to maintain a fashionable wardrobe, and the desire to be eco-friendly [41].

We see that the *sharing economy* is one business model (and slogan) to fight against what we call *consumerism*. We will return to the sharing economy model at Sect. 6.2. But what are we speaking about when we say "consumerism"?

1.2.5   Consumerism

The changes created by the Industrial Revolution and the rise of mass production during the twentieth century helped the United States become a producer society, and rising standards of living also created a hungry class of consumers interested in purchasing those goods. The United States increasingly consumed goods produced elsewhere and outsourced the pollution and human costs of mass production to other countries, while its own demand for products continued to increase.

In order to increase consumption and the motivation of people to buy and buy and buy, new market strategies were invented to persuade people to acquire products they don't necessarily need. Vance Packard's commemorated best seller, *The Hidden Persuaders*, published in 1957, warned Americans that advertisers use psychological techniques to manipulate people to desire products that companies want to sell [42]. One of the keywords in the book was "unconscious." The advertising industry helped sell people things they don't need and generated consumerism.

The wardrobe of the heroine of the popular series *The Marvelous Mrs. Maisel*, who went from an ordinary housewife to an extraordinary comedienne, reflects well the world of upper-middle-class Manhattan in the late 1950s. The costumes of the celebrated designer Donna Zakowska—from skirt suits to velvet hats to evening gloves—fit well with her luxury, uptown family life. She changes her outfit in the late evening to fit with the bohemian spirit of the downtown comedy clubs in which she performs. Throughout the show, we see her extensive wardrobe, whose abundance mirrors well the atmosphere of consumerism.

Consumerism has been a double-edge sword. It would be difficult to deny that consumerism is a driving force of the economy. We have known for decades that it destroys the environment, and it is often based on the exploitation of people, including through child labor, poorly paid work, and unsafe working conditions. Consumerism is a controversial topic, with both pros and cons [43].

Pros:

1. Consumerism stimulates economic growth.
2. It also boosts creativity and innovation.
3. Cost reductions are encouraged because of consumerism.
4. It weeds out the poor performers naturally.
5. Consumerism encourages freelancing, entrepreneurialism, and self-employment.
6. It creates safer goods for consumers.
7. Consumers are given more choices in this society.

Cons:

1. The economy takes precedence over the environment.
2. It changes the moral fabric of society.
3. Consumerism encourages debt.
4. It leads to health problems.
5. Consumerism does not provide fulfillment.
6. It can be used as a political tool.
7. Consumerism conflicts with various spiritual beliefs.
8. The poor are always left behind by consumerism.

As an illustration, we mention here a paradoxical relationship between consumerism and tradition. Japanese people passionately wrap up literally everything, from small gifts and goods to cars. It helps sell things, but it also generates a lot of waste. The traditional concept, known as "tsutsumu," uses natural materials like bamboo, ceramics, leaves, and rice straws. Objects are conscientiously wrapped and bagged in multiple layers of packaging material. Convenience stores (konbini) are very important locations of the contemporary Japanese life. In 2019, the country had 56,502 such shops. Characteristically, they are open 24 h per day, 365 days per year [44]. They sell ready-made food for immediate consumption (and offer other services from photocopying to money transfers). The take-away culture of the konbini correlates with plastic consumption. Employees have always carefully packed items into carrier bags, making sure that nothing will be damaged. Warm and cold items are separated by bags. Items that can leak are packed even better. They are individually wrapped in another plastic bag beforehand. In addition, free single-use cups, cutlery, chopsticks, and napkins are automatically provided. Slowly, we have heard calls for an end to throw-away society. In the summer of 2020, Japanese retail shops began charging customers for plastic bags. The rule emerged as a response to domestic and international pressure to reduce plastic bag use. As the rule came synchronously with the emergence of COVID-19, people felt that single-use packaging is safer, since no one has touched the item before. With the introduction of the levy on the bag—even if it is very small but larger than zero—it means that it is no longer a free gift. The merchant has to ask whether the customer wants to buy one or not. Due to different aspects of the traditional Japanese shyness, customers don't prefer to refuse bags.

1.2.6 The Throw-Away Society: Emergence and Call for the End

Individual disposable products emerged by design. Personally, we can identify the transition from reusable diapers to disposable ones. Péter enthusiastically ironed diapers in 1976 and in 1982. Zsuzsa's second and third children were born a decade apart, so Csilla, the youngest of the five kids in the two families was the only "disposable diaper" baby, while poor Gábor, Zsuzsi, Zsófi and Miklós grew up with ironed ones. We should provide a fair assessment of the ecological footprint of both methods (i.e., use of reusable or disposable diapers) before making general statements. Reusable diapers have to be washed and ironed. Washing produces wastewater containing detergents, and ironing needs electricity, the production of which also generates waste.

Let's turn back from the very personal to the global: The throw-away society emerged as a consequence of economic growth. A number of factors can be identified, as the sources of "waste":

- The short life span of products requires consumers to repurchase goods frequently;
- Problems with irreparable goods encourage the discarding of objects too early;
- Excessive packaging constitutes its own form of waste;
- Disposable goods from tableware provided by fast food chains to fast fashion clothes have become very popular.

A New Slogan: From a Linear Economy to a Circular Economy

Figure 1.1 illustrates in a condensed form the difference between the models of linear and circular economies.

The *linear economy* is the traditional model based on "take-make-consume-waste" approach to using resources. Resources, as raw materials, are transformed into a product. Each product has a life-span, and after that life-span ends, the products are thrown away. It looks plausible that the model should be replaced to maintain a more sustainable world.

The transition from the linear to the circular economy emphasizes the concept of *reuse*. Products are supposed to be recycled into a new product when they end their life cycles. The main goal of the circular economy is to prevent waste. Manufacturers should design reusable products, and when the raw material enters the cycle, it must follow the path of "production-use-recycling-production."

At least in principle, the circular economy might be both more profitable and less harmful to the environment, and its main goals include sustainable economic growth. Please note, we wrote "might be." We should conduct an ecological footprint

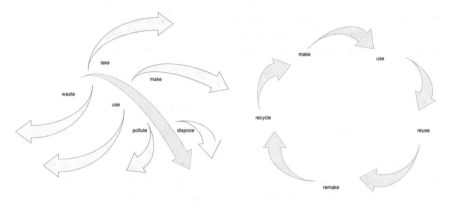

Fig. 1.1 Linear economy versus circular economy. *Adapted from Ref.* 1.1, p. 18

analysis in each case before deciding in favor of recycling. In many cases, it is not trivially true that remaking or recycling is more environmentally friendly and more profitable than single-use products and safe waste management.

More emphatically, the "out with the old, in with the new" mentality implements a lifestyle that is enormously costly for both personal finance and for the environment. As the financial journalist Lisa Smith writes:

> Get Started Today
> "Ignore the siren song of runaway spending. Forget about owning the latest styles, biggest houses or flashiest cars. Instead, make your financial situation your top priority. Your pocketbook will breathe a sigh of relief when you make the effort—and you may be able to reduce your impact on the environment in the process" [45].

Among others, Japan established the Japan Partnership for Circular Economy with the goal of "strengthening public and private partnerships, with the aim of further fostering understanding of the circular economy among a wide range of stakeholders, including domestic companies, and promoting initiatives in response to the accelerating global trend toward a circular economy" [46].

1.2.6.1 The Replace or Repair Dichotomy: Toward a New Perspective on Resource Management

Many of us (we guess it is still a minority) feel that having more stuff does not necessarily make us happier. We are facing now what we might call the "replace or repair" dichotomy.

First of all: When should I repair or replace a good? The technical answer is not too complicated: The current value of the asset should be compared with the cost of repair. When the cost of repair is less than than the value of the piece, you should repair it. Otherwise, you should replace it. We offer some not-very-scientific data about our attitude: If the repair costs exceed about a third of the price of the new item, we incline toward buying a new one. There is a more spiritual answer, as we all know: "Lord, give me strength to change the changeable, patience to endure the immutable, and wisdom to distinguish the two!"

Second, repair is not always easy. The logic of the planned obsolescence naturally led to the formation of a business model that forbids or makes it very difficult to repair a device by substituting a relatively cheap component. Sometimes components are glued in place, which makes them difficult to remove. Apple, for example, adopted a technology that made it impossible to repair their products unless done by the company's own technicians.

The *right to repair* movement is a very important step toward normalizing the practice of repair. The goal of the movement is to encourage companies to make spare parts, tools, and information on how to repair devices available to customers, to increase the lifespan of products, and to prevent them from ending up in landfills. The capsule history of the right to repair movement will be reviewed in Sect. 4.3.

Third, to make the transition to a repair society, people should realize that not only are goods and gadgets resources, but we also have human resources, such as family members, friends, and the small groups and large communities to which we belong. One of our main points in this book is an attempt to explain how we should manage *both* our physical and social resources. So, we should discuss now our most important resources, our *social relationships*.

1.3 Social Relationships as Resources

How many people did you talk to yesterday? How many people can you tell that you had a fight with your lover? How many people can you sit down for a beer with? There are different layers of friendships and other social relationships, and we need all types of engagement to live healthy social lives.

1.3.1 Friends Versus Acquaintances

How many friends do you have? Despite what Facebook might suggest, we cannot have 1000 friends. We cannot even have 1000 close acquaintances. The British anthropologist Robin Dunbar has estimated that the number of persons with whom we can form stable social relationships is approximately 150, which more precisely means between 100 and 200. Please note, this number applies to quality relationships, not to acquaintances. Social resources matter—we should not assume that only physical resources talk.

Humans like paradoxes. The most cited paper in *Sociology*, published by Mark Granovetter in 1973, analyzed one such paradox [47]. In certain situations weak connections are more efficient than strong ones. With some simplification, we have both strong and weak ties. Families, friends, and close colleagues are strong ties, and more distant acquaintances with whom we meet and communicate infrequently are weak ties.

Why may a weak connection prove more efficient? The main idea is that people with solid connections form a clique. Everybody knows everybody, and they share knowledge and background. As a result, they will *not* produce novelty. However, weak ties might play the role of a *bridge* by connecting different communities, and they might be a channel for the flow of new information from other communities. Specifically, Granovetter found that weak ties played a significant role in finding new jobs. (His data came mainly from the 1960s. Social media's impact on job seeking and applying has since changed the landscape.) The problem is that many people apply for the same job, and the question we face is how to be selected (or, if you wish, how to play the "ranking game" [48]). Human Resources managers have limited tools to find the "best candidates" among hundreds of applications. A single strong recommendation from a reliable source might be a decisive factor. A reliable source might, for example, be a colleague from your previous job. So, if you have a good connection with them when you are in the same workplace, you may have a better chance of finding a new one thanks to their recommendation [49].

The Evolution of Friendships

Close friendships emerged in early civilizations among contemporary hunter-gatherers. People had to cooperate to hunt mammoths, so the friendship was of a *physical* character. The role of friendship has since evolved from physical to psychological: "We do things for the other out of friendship, not in order to gain anything. Friendships can provide grounding, safety, comfort, the experience of trust and respect, of being understood and valued" [50]. Historically, we needed friends for survival. We have institutions, ideally from the police to the state, to ensure survival. (We wrote "ideally"; real police often provide control and discipline). Still, we need friends to help us navigate the psychological aspects of life. We need people with whom to to celebrate in good times and from whom to receive consolation in bad times.

The cognitive anthropologist Brian Hare argued that our ability to make friendships—otherwise speaking, to see evolution more as cooperation and less as competition—helped us survive. While we shared the planet with at least four other types of humans, it turned out that not aggression but the ability to cooperate was the most important trait for survival [51].

There is a frequent dynamic pattern in the modern world: We have lots of friends in middle and high school, maybe even more at university. Then …we lose them one by one. We have heard recent stories that friendships among former high schoolmates reemerged decades after a break. Broken friendships can be repaired. Or not.

Jean De La Bruyère (1645–1696), a French philosopher and moralist wrote: "Two people cannot be friends for long if they cannot forgive each other for their little faults." Ralph Waldo Emerson (1803–1882), an American essayist, highlights the importance of prevention: "Keep your friendships in repair." Samuel Johnson (1709–1784), an English writer, further emphasizes prevention: "A man, sir, should keep his friendship in constant repair."

Some of our broken friendships may never be recovered, but some of them are worth repairing. Our lives are enriched by the experiences they provide, and they make us better and happier. But how do we know if a friendship is worth it? It costs us a lot to have a friend who doesn't respect our principles, who forces us to compromise on our beliefs, or who makes us lose self-esteem. We may tell (or write) the other person clearly and directly that our friendship is important to us but that a conflict, a misunderstanding, or simply too much time without their company makes us feel we have to move on. Without complaining or listing grievances, we may also ask our friends how they see the situation and whether they think there is a way to resolve it. We may then decide that there is no room for accountability, revenge, or punishment, even if "we are right." Or we may forgive the other person if we have been looking for them and they have not been in touch for months. Then, if a meeting takes place and the problem cannot be clarified and discussed sufficiently, it may not be worth continuing the friendship. Sometimes it's a relief when things are finally resolved.

Respect and responsibility play an important role in a new beginning. It is more efficient than to feel guilt. Our friendships can be restored, reconsidered, or let go clearly and unequivocally. We can only come to terms with each of these versions of the future if we try them [52].

While it is very important to develop strong personal relationships with individuals, we are members of several social groups—physical and virtual communities. They are valuable resources for us.

1.3.2 Social Groups as Resources

Evolutionary Origins of Social Groups

Integrated activity enhances the power of individuals to achieve their goals. It is fruitful to be a member of a group. It is well known that many species of animals (like insects, fish, and wolves) live in groups to increase their chances of survival. Edward Wilson (1929–2021) explained altruism, aggression, and other social behaviors in biological evolution. His book on what he called *sociobiology*, which mainly dealt with social animals like ants, included a single chapter on humans and provoked sharp debates [53]. The opponents of sociobiology were headed by leading (admittedly leftist) evolutionary biologists, Richard Lewontin (1929–2021) and Stephen Jay Gould (1941–2002) who attacked sociobiology for supporting biological determinism. Biological determinism may have, as they argued, serious negative social consequences. Sociobiology has been replaced by *evolutionary psychology*, a less direct, more neutral theory that explains the evolution of human behavior and culture through the mechanism of natural selection [54]. Evolutionary psychology likes to see many human behaviors in animals. Zebras, antelopes, bison, and several other herbivores have the ability to spontaneously form "gangs" when predators appear because herds are more likely to escape. It was an infamous story in 2017 when a crazed zebra gang attacked a young antelope in front of shocked tourists [55].

Okay, but we are humans …

Our Role in Multiple and Dynamic Groups

People form *groups*—formal ones like organizations and informal ones like friend groups. Groups in a gym for exercise classes are semi-formal. There are more enduring groups, and there are more transient ones. Classical social psychology concentrates on the behavioral patterns of well-defined, static groups. While this approach certainly had successes, we now see that (i) the dynamical systems approach to social psychology emphasizes the transient, dynamic behavior of groups [56]; and (ii) people belong to several different groups and have diverse functional roles in these groups. For example, a person may belong to an ethnic and/or religious group, as well as groups of colleagues, classmates, sports teams, etc. The subjective feeling of belonging to these groups varies and changes. The success of the group activity may strengthen the sense of belonging.

Here is an example, and everybody knows many similar ones.

Zsuzsa has a younger, well-socialized friend, let's call her Mary. She is a 34-year-old assistant at a financial institution and has a leading role in the local sports club in the afternoon. It takes a lot of time and effort, but she knows almost everyone. Many families are connected to the sports club because of their children.

She takes care of the cookies as a member of a congregation on weekends. For Mary, it's a much less intense role, with fewer people to meet, and these encounters are more controlled and shorter. She also administers the collection of secondhand clothes in the congregation, so she has multiple functions in the same group. But this only takes an hour or two a month, and most donors are contacted by email.

Mary also started learning Japanese in an adult group organized in her neighborhood a year and a half ago, and she goes to play the clarinet in the band of the neighboring village because the clarinetist there unexpectedly went abroad. The first group only learns language, the second only music—the joint activities are a distinctive feature of the content and nature of the time spent together.

Mary is a member of many groups and plays countless roles and functions in these communities. But she is attached to groups and relationships in different ways, using different amounts and qualities of resources to maintain these relationships.

There are different synchronization mechanisms to organize interpersonal, group, or large-scale activities. Compassion and empathy help synchronize emotions like joy, sadness, and worry [57]. While we like to think that empathy is a purely human trait, dogs, elephants, and apes, among others, are also able to express complex emotions like empathy.

Consensus versus *Echo Chambers*: A Dichotomy

Initial differences in opinions and attitudes of group members are very natural. The process of coordination based on communication and actions to reduce the discrepancies between individual attitudes is vital in any group solving task. We have tools like eye contact, facial expressions, body language, and words to shape our shared goals: We are ready to synchronize. Synchronization often leads to *consensus* and is an essential form of coordination that increases our feeling of belonging. Lack of synchronization can contribute to a sense of isolation and loneliness. There are different mechanisms for syncing: It occurs in dyads, among several people, in small communities, and larger groups. Homophily, "love of sameness," might be a sociological mechanism behind group formation.

Decision making with consensus: Why do we like it, and why don't we?

The writers and the majority of Readers might share a belief that consensus has many beneficial aspects in decision making. Discussions on differences might lead to shared understanding. Discussion helps keep the distribution of power in the group balanced and increases the responsibility of the participants. To remain positive, we would like to believe that better decisions are being brought by consensus.

But many times, we are simply wrong. Should we admit that consensus is not necessarily always good? In modern democratic societies, we naturally consider shared decision making superior to autocracy. The accumulation of financial and political power in the hands of the few has frequently led to corruption and to very biased decision making. However, "singular decision power given to the right person, at the right time, in the right amount, is one of our very most effective tools" [58]. The key is the existence of institutional control over the activity of this person.

However, people in a group have different backgrounds, knowledge, expertise, and skills. It can be more effective if decision making is delegated to members with higher competence. A pilot, not the consensus, should drive a plane. This might be true for companies, too. We all know the saying: "A camel is a horse designed by a committee," as well as the old idiom "too many cooks spoil the broth." The superiority of diverse groups in problem solving will be discussed a little later. But now, let's study some opposing views related to too much homogeneity.

Like it or not, an unintended consequence of the use of social media is that it seems more and more challenging to reach consensus. Online political discussions are prevalent. Since internet technology allows for the efficient use of filtering, we communicate primarily in *echo chambers* among people with similar ideological principles. We receive predetermined information delivered in a personalized journal, the "Daily Me." These *echo chambers* are means for amplifying our beliefs [59]. (Positive feedback is a primary self-amplifying mechanism, which we will discuss in Sect. 3.3.2.) Communication occurring in isolated parallel channels is dangerous and may contribute to political polarization. Group polarization makes cross-talk almost impossible. We return to discuss the mechanisms of group dissolution in Sect. 3.4.4.

Diversity

Péter's student Caroline Skalla studied diversity in an agent-based computational model [60]. We take the liberty to use some parts of her undergraduate thesis almost verbatim. (Thank you, Caroline!)

Diversity is a term that comes up in many conversations and can have various meanings. However, the most common references to diversity usually pertain to demographic diversity, which is often measured by characteristics such as age, race, or ethnicity [61].

An encompassing definition of diversity is below:

Diversity describes the distribution of differences among the members of a unit concerning a common attribute, X, such as tenure, ethnicity, conscientiousness, task attitude, or pay [62].

The quality measured is context-dependent. For example, if socioeconomic diversity was the context of interest, one might use attributes such as gender, age, ethnicity, education level, or income to characterize the differences among a population.

Social groups are assets for individuals. What is a reasonable expectation about the composition of a group? Intuitively, we feel that too much homogeneity is unfavorable, since it does not allow for any flexibility. Too much heterogeneity also does not seem to function well, since endless debates on the most basic principles cannot be productive. But here is a question: Is there an optimal composition for a social group faced with some task? We give some specific answers for this question.

Hedgehogs and Foxes: In Praise of Diversity During Problem Solving

The Greek poet Archilochus (c. 680–645 BCE) famously wrote that "The fox knows many things, but the hedgehog knows one big thing." The philosopher Isaiah Berlin (1909–1997) combined his Russian Jewish roots with his education to become a central member of Oxford's philosophy school (and, arguably, of the whole English intellectual society). He adopted the idea in his 1953 essay *The Hedgehog and the Fox* (about Tolstoy's view of history). He contrasted the hedgehogs who "relate everything to a single central vision" with foxes who "pursue many ends connected …if at all, only in some de facto way."

There are two types of people. One focuses on a single big idea on which they always rely. The other one takes a fresh approach to every problem. (While this ancient philosophy was written over 2000 years ago, it fits surprisingly well with modern career advice. The internet is plastered with articles advertising the best way to boost one's career, citing the wisdom of the ancient Greek philosophy: "Are you a fox or a hedgehog? Here's what an old saying reveals about your leadership.")

Robust research suggests that generalists are better at navigating uncertainty. Philip Tetlock, a professor of Organizational Behavior at the University of Pennsylvania, has found experts are less accurate predictors than non-experts in their areas of expertise [63]. Hedgehogs were very confident in their predictions because they believed their "Single Big Idea" empowered them to understand the reality they perceived only superficially. However, their performance was often worse than random guessing! Foxes have a better sense of fact, they don't believe in the omnipotence of

Big Theories, and they perceive that the world's complexity is determined by many causal factors. While they were less confident in their predictions, their performance still exceeded the results of random guessing.

Foxes and hedgehogs might sound like just a cute analogy. Still, this proverb was an early observation of cognitive diversity—the way people think and approach problems—which has become a topic of great scientific interest in the present day. The differences in how people encode, store, and process information is relevant to nearly every field, from neuroscience to education to policymaking. Studying these cognitive differences is challenging and requires expertise from various disciplines. However, the world is growing interconnected, so learning about cognitive differences among individuals is insufficient. Instead, we must explore how groups of cognitively diverse individuals interact. Governments, juries, research groups, and corporations…

Combined observational and computational studies demonstrate that teams solve problems better or faster if they show *cognitive diversity* [64]. Cognitive diversity has been defined as differences in perspectives or information-processing styles and does not necessarily correlate with gender, ethnicity, or age.

Hong and Page, for example, found that moderately intelligent but heterogeneous groups outperformed the most intelligent but homogeneous agents [65]. But, we have a but: It goes back to the concept of *groupthink*, a term coined by the Yale University social psychologist Irving Janis (1918–1990) in 1972 [66]. Janis found that groups of intelligent people sometimes make the worst possible decisions. One reason was that the group members might all have similar backgrounds or and be isolated from the opinions of other groups. They also believe that they are right and that opinions from other sources could be neglected. It is now believed that huge disasters, such as the explosion of the space shuttle Challenger in 1986, and events like the Bay of Pigs invasion, Watergate, and the escalation of the Vietnam War emerged from uncontrolled groupthink. While the Hong and Page paper's message is still crucial, Grim and his coworkers advise that these results should not be over-interpreted when making policy [67].

Caroline studied the ratio of foxes to hedgehogs a team needs to solve the greatest number of tasks in a cooperative, self-organized environment [67]. Her simulation studies support the view that functionally diverse teams perform the best when *information sharing* increases.

Repairing Friendships: Some Preview

To put it sharply, certain things and relationships can be improved. Others cannot. It is not always easy to discriminate between the two possible outcomes. Those who learned early in life that things should not be lost are primed to maintain things and relationships. To preserve what we still think belongs to us, even if it is not as it was when it was new.

Even if we repair the torn bag strap for the fifth time, we could have learned the lesson after the third time that it was simply of a lousy quality. Even if we decided that *we* would be the ones to call our old friend regularly because *we* are the one who doesn't want the relationship to weaken, the question of whether it is worth saving something can arise at any time. To answer this question, it is worth thinking about (i) the lack of a particular thing, (ii) its substitutability, and (iii) the resources spent on replacement.

As the psychiatrist Irvin D. Yalom writes, we will never have new old friends [68]. No carpentry workshop makes an oak-table on which Russian soldiers burned the children's books of Zsuzsa's mother. In addition, the process of repairing can itself can be rewarding. Minutes of sewing a torn sweater can often provide an escape from the unpleasant noises of the big, accelerated world.

Sometimes we are too needy or expect too much without giving enough back in return. So be cautious. It is heartbreaking to lose good friends. We will return to discuss the patterns of damaged relationships in Sect. 3.4.

1.4   Lessons Learned and Looking Forward

We started the book by giving hints about our motivations for repair. Albeit separated by a generation, we still grew up in a relatively poor country, packed with memories of a war preserved in both physical structures and souls. So we believe we have an instinct not to waste.

Despite increasing inequality, the developed world now lives with the blessing and curse of the throw-away society. In the beginning, the benefits seemed to vastly exceed the costs, and the broadly defined middle class enjoyed the fruits of consumerism. Then, however, what first appeared to be only collateral damage (e.g., food and clothing waste or dramatic increases in pollution), became the predominant feature of our society. While it is difficult to change the self-driving mechanisms of global capitalism, more and more voices shout that the smooth continuation of the present trajectory may lead to ecological, economic, political, and social crises. Therefore, the throw-away society should make fast and responsible decisions to transition to what we may call a repair society.

One of our main points is that we need to manage our social relationships, both individually and in groups. We need mates and friends, and we should belong to several different groups to maintain our working life by solving tasks and problems and preserving our emotional stability.

We can repair things that were first functioning well but were impaired by some mechanism. To take one more step back, we may find ourselves in the probably never existing past, which we might call a Golden Age. So, we now make an excursion to the universe of peace, where nature and people were supposed to live in harmony. Did they?

References

1. Molnár V (2018) Utopian visions in the rubble: constructing a New City versus reconstructing the old in post-war budapest. In: Proceedings of the symposium on post war reconstruction: the lessons of Europe. Lebanese American University School of Architecture and Design. http://sardassets.lau.edu.lb/arc_catalogs/post-war-reconstruction/post-war-reconstruction-7.pdf
2. Dryer G (2020) Letter of recommendation: rags. The New York times magazine. https://www.nytimes.com/2020/02/19/magazine/letter-of-recommendation-rags.html
3. Hazlitt H (1996) Conquest of poverty. Foundation for economic education. https://fee.org/resources/the-conquest-of-poverty/
4. Alfani G, Gráda CÓ (2018) The timing and causes of famines in Europe. Nat Sustain 1:283–288. https://doi.org/10.1038/s41893-018-0078-0
5. Malthus T (1798) An essay on the principle of population as it affects the future improvement of society, with remarks on the speculations of Mr. Goodwin, M. Condorcet, and Other Writers. J. Johnson in St. Paul's Church-yard. London. http://name.umdl.umich.edu/004860797.0001.000
6. Malthus, the false prophet (2008). The Economist. https://www.economist.com/finance-and-economics/2008/05/15/malthus-the-false-prophet
7. Tabarrok A (2003) Productivity and unemployment. Marginal Revolution. https://marginalrevolution.com/marginalrevolution/2003/12/productivity_an.html
8. Roodman D (2020) Modeling the human trajectory. Open philanthrophy. https://www.openphilanthropy.org/blog/modeling-human-trajectory
9. Acemoglu D, Restrepo P (2018) Modeling automation. AEA papers and proceedings 108:48–53. https://doi.org/10.1257/pandp.20181020
10. Pethokoukis J (2018) Yes, AI can create more jobs than it destroys. Here's how. American Enterprise Institute. https://www.aei.org/economics/yes-ai-can-create-more-jobs-than-it-destroys-heres-how/
11. Naso P, Lanz B, Swanson T (2020) The return of Malthus? Resource constraints in an era of declining population growth. Eur Econom Rev 128:103499. https://doi.org/10.1016/j.euroecorev.2020.103499
12. Friedman T (2005) The world is flat. Farrar, Straus and Giroux
13. Pinker S (2018) Enlightenment now: the case for reason. Science, Humanism, and Progress. Viking
14. Kenton W (2019) Planned obsolescence. Investopedia. https://www.investopedia.com/terms/p/planned_obsolescence.asp
15. The 1950s Analysis. Shmoop. https://www.shmoop.com/study-guides/history/1950s/analysis#economy
16. Krajewski M (2014) The great lightbulb conspiracy. IEEE Spectrum. https://spectrum.ieee.org/the-great-lightbulb-conspiracy
17. Hadhazy A (2016) Here's the truth about the 'planned obsolescence' of tech. BBC Future. https://www.bbc.com/future/article/20160612-heres-the-truth-about-the-planned-obsolescence-of-tech
18. Forti V, Balde CP, Kuehr R, Bel G (2020) The global E-waste monitor 2020: quantities, flows and the circular economy potential. United Nations University/United Nations Institute for Training and Research, International Telecommunication Union, & International Solid Waste Association. https://collections.unu.edu/view/UNU:7737
19. Fast Facts About Agriculture and Food. American Farm Bureau Federation. https://www.fb.org/newsroom/fast-facts
20. McCarthy L (2020) When hunger is on the doorstep. New York Times. https://www.nytimes.com/2020/09/02/insider/food-insecurity-families.html
21. China launches 'Clean Plate' campaign against food waste (2020) BBC News. https://www.bbc.com/news/world-asia-china-53761295

22. Marchiso M (2020) Fighting food waste in China: local efforts, global effects. International Fund for Agricultural Development. https://www.ifad.org/en/web/latest/-/blog/fighting-food-waste-in-china-local-efforts-global-effects
23. Singh A, Sarda A (2020) Food wastage in Indian weddings. Don't Waste My Energy! https://dontwastemy.energy/2020/03/24/food-wastage-in-indian-weddings/
24. Chandler A (2016) Why Americans lead the world in food waste. The Atlantic. https://www.theatlantic.com/business/archive/2016/07/american-food-waste/491513/
25. Klein C (2021) Mobster Al capone ran a soup kitchen during the great depression. History. https://www.history.com/news/al-capone-great-depression-soup-kitchen
26. Roy P (2019) The 9 best food waste apps to make sustainable eating easier. Vogue. https://www.vogue.co.uk/gallery/best-food-waste-apps
27. MICHELIN Green Star restaurants in Japan (Accessed 5 Sept 2021). MICHELIN guide. https://guide.michelin.com/en/jp/restaurants/sustainable_gastronomy
28. The Nobel Peace Prize 2020. (Accessed 17 Apr 2022). NobelPrize.org. https://www.nobelprize.org/prizes/peace/2020/summary/
29. 28 Types of Fabrics and Their Uses (2021). MasterClass. https://www.masterclass.com/articles/28-types-of-fabrics-and-their-uses#28-different-types-of-fabric
30. The Problem With Plastics (2022) Ocean conservancy. https://oceanconservancy.org/trash-free-seas/plastics-in-the-ocean/
31. Thompson A (2018) From fish to humans, a microplastic invasion may be taking a toll. Sci Am. https://www.scientificamerican.com/article/from-fish-to-humans-a-microplastic-invasion-may-be-taking-a-toll/
32. Jambeck JR, Geyer R, Wilcox C, Siegler TR, Perryman M, Andrady A, Narayan R, Law KL (2015) Plastic waste inputs from land into the ocean. Science 347(6223):768–771. https://doi.org/10.1126/science.1260352
33. Resnick B (2019) More than ever, our clothes are made of plastic. Just washing them can pollute the oceans. Vox. https://www.vox.com/the-goods/2018/9/19/17800654/clothes-plastic-pollution-polyester-washing-machine
34. Borelle S, Ringma J, Law KL, Monnahan CC, Lebreton L, McGivern A, Murphy E, Jambeck J, Leonard G, Hilleary M, Eriksen M, Possingham HP, de Frond H, Gerber LR, Polidoro B, Tahir A, Bernard M, Mallos N, Barnes M, Rochman CM (2020) Predicted growth in plastic waste exceeds efforts to mitigate plastic pollution. Science 369(6510):1515–1518. https://doi.org/10.1126/science.aba3656
35. Butler S (2019). Why are wages so low for garment workers in Bangladesh? The Guardian. https://www.theguardian.com/business/2019/jan/21/low-wages-garment-workers-bangladesh-analysis
36. Jacoby J (2018) What's changed (and What Hasn't) since the Rana Plaza Nightmare. Open Society Foundations. https://www.opensocietyfoundations.org/voices/what-s-changed-and-what-hasn-t-rana-plaza-nightmare
37. Thomas D (2020) Fashionopolis: why what we wear matters. Penguin Books
38. Wood K (2022) 8 Reasons to rethink fast fashion. Lifehack. https://www.lifehack.org/articles/money/8-reasons-rethink-fast-fashion.html
39. From old to new with Looop (2020) H&M. https://www2.hm.com/en_gb/life/culture/inside-h-m/meet-the-machine-turning-old-into-new.html
40. Park H, Martinez CMJ (2020) Secondhand clothing sales are booming—and may help solve the sustainability crisis in the fashion industry. The Conversation. https://theconversation.com/secondhand-clothing-sales-are-booming-and-may-help-solve-the-sustainability-crisis-in-the-fashion-industry-148403?
41. Taelor (2022) F6S. https://www.f6s.com/taelor
42. Packard V (1957) Hidden Persuaders. Ig Publishing
43. Gaille L (2018) 15 Consumerism Pros and Cons. Vitanna.org. https://vittana.org/15-consumerism-pros-and-cons

44. Steger B (2021) Stingy, stingy, stingy government: mixed responses to the introduction of the plastic carrier bag levy in Japan. Worlwide Waste 4(1):5. https://www.worldwidewastejournal. com/articles/10.5334/wwwj.69/
45. Smith L (2021) The disposable society: an expensive place to live. investopedia. https://www. investopedia.com/articles/pf/07/disposablesociety.asp
46. About the Japan Partnership for Circular Economy (Accessed 17 Apr 2022) Japan Partnership for circular economy. https://j4ce.env.go.jp/en/about
47. Granovetter M (1973) The strength of weak ties. Am J Soc 78(6):1360-1380. https://www. jstor.org/stable/2776392
48. Érdi P (2019) Ranking: the unwritten rules of the social game we all play. Oxford University Press
49. Gershon I (2017) A friend of a friend" Is no longer the best way to find a job. Harvard Buiness Rev. https://hbr.org/2017/06/a-friend-of-a-friend-is-no-longer-the-best-way-to-find-a-job
50. Bell P (2014) The history of friendship: how friendship evolved and why it's fundamental to your happiness. Huffington Post. https://www.huffingtonpost.co.uk/2014/02/10/history-of-friendship-evolution_n_4743572.html
51. Hare B, Woods V (2021) Survival of the friendliest: understanding our origins and rediscovering our common humanity. Random House
52. Parrott L, Parrott L (2002) Relationships: how to make bad relationships better and good relationships great. Zondervan
53. Wilson E (1975) Sociobiology: the new synthesis. Belknap Press
54. Barkow J, Cosmides L, Tooby J (1995) The adapted mind: evolutionary psychology and the generation of culture. Oxford University Press
55. McDonald A (2017) Crazed zebra gang attack young antelope in front of shocked tourists. Daily Star. https://www.dailystar.co.uk/news/latest-news/zebra-attack-gang-video-fight-17014132
56. Nowak M, Sigmund K (2005) Evolution of indirect reciprocity. Nature 437:1291–1298. https:// www.nature.com/articles/nature04131
57. Provine RR (1992) Contagious laughter: laughter is a sufficient stimulus for laughs and smiles. Bullet Psychonom Soc 30(1):1–4. https://doi.org/10.3758/BF03330380
58. Peshawaria R (2017) Is consensus always a good thing? Forbes. https://www.forbes.com/sites/ rajeevpeshawaria/2017/10/29/is-consensus-always-a-good-thing/?sh=1c11be144c43
59. Sunstein CR (2017) #Republic: divided democracy in the age of Social media. Princeton University Press
60. Skalla C (2021/2022) Modeling cognitive diversity. Senior Individualized Project. Kalamazoo College
61. Mansoor S, French E, Ali M (2020) Demographic diversity, processes and outcomes: an integrated multilevel framework. Manage Res Rev 43(5):521-543. https://doi.org/10.1108/MRR-10-2018-0410
62. Harrison DA, Klein KJ (2007) What's the difference? diversity constructs as separation, variety, or disparity in organizations. Acad Manage Rev 32(4):1199–1228. https://doi.org/10.5465/amr. 2007.26586096
63. Tetlock PE (2006) Expert political judgment: how good is it? how can we know? Princeton University Press
64. Reynolds A, Lewis D (2017) Teams solve problems faster when they're more cognitively diverse. Harvard Business Rev https://hbr.org/2017/03/teams-solve-problems-faster-when-theyre-more-cognitively-diverse
65. Hong L, Page SE (2004) Groups of diverse problem solvers can outperform groups of high-ability problem solvers. Proc Nat Acad Sci USA 101(46):16385–16389. https://doi.org/10. 1073/pnas.0403723101
66. Janis IL (1972) Victims of groupthink: a psychological study of foreign-policy decisions and fiascoes. Houghton Mifflin
67. Grim P, Singer D, Bramson A, Holman B, McGeehan S, Berger W (2019) Diversity, ability, and expertise in epistemic communities. Philosophy Sci 86(1):98–123. https://www.journals. uchicago.edu/doi/epdf/10.1086/701070
68. Yalom ID (2009) Staring at the sun: overcoming the terror of death. Jossey-Bass

Chapter 2
A Golden Age that Never Was

Abstract In this chapter, we review the concept of the Golden Age, the mythical era when everyone lived in peace, harmony, stability, and prosperity. People often have an unconscious instinct and conscious idea that there was once a perfect state of existence, both at the individual and collective levels. The loss of Paradise due to expulsion from the Garden of Eden remains unprocessed despite the continuous reflections of artists in changing times. The Greco-Roman and the various Asian golden ages, and their falls, are a dominant part of human history. In the life histories of individuals, the search for and knowledge of the specific period of true bliss goes back to the beginning of childhood. In social history, it stretches back to ancient sources. The combined effect of individual and historical memory leaves many of us feeling that there is always something to repair or improve in our life, in its various dimensions.

2.1 The Myth of the Golden Age: Looking into the Past

We should start the central part of the book with a discussion of the "Age Before Things Went Wrong." Why? Elementary logic requires that if things must now be repaired, then at some time in the past they were superior, and perhaps we could return to this state in the future. During the moments (months and years) of crisis, we like to believe that we will be able to return to the abundance and peace that humankind might have had in the Golden Age. However, one with a cynical nature might suggest that we are thinking about a Golden Age that never was.

In this chapter, we review the concept of the Golden Age, the mythical era when everyone lived in peace, harmony, stability, and prosperity. We use the term "harmony" here to mean a condition in which people have similar views about things, or at least views that do not lead to massive, sustained conflict with other people and their communities. People often have an unconscious instinct and conscious ideas that there was once a perfect state of existence, both in our own lives, at the individual level, and collectively, in social groups, too. In the life histories of individuals, the search for and knowledge of the period of true paradise goes back to childhood. For many of us, it seems that elementary school was the golden age of childhood (this idea is also found in the literature on child psychiatry [1]). In social history, this belief in a golden age stretches back to ancient sources. The combined effect of individual and historical memory leaves many of us feeling that there is always something to repair or improve in our lives, in their various dimensions. As a result, we often desire to return to an earlier state, which may or may not have existed.

During the whole of human history, people have been fighting for survival in a threatening world. Since the beginning, we have used our imagination to try to understand life on our planet, which sometimes looks cruel, sometimes generous. One way to cope with the difficulties is to tell ourselves stories that give meaning to larger-scale changes. Myths suggest that we come from a period of peace: a time when the seasons did not change; people were not threatened by cold, famine, and loneliness; and the gods still loved them. We use our imagination to try to regain the original sense of belonging we believe we once had. Ancient myths, at least as old as those of the Greek poet Hesiod, have tried to help us understand our place in the world. We all want to know where we came from. Since our earliest beginnings are lost in the mists of distant prehistory, we have created myths about our forefathers that are not necessarily historical but help to explain current attitudes about our environment, other people, and customs. Please note, myths are not *fake news* but frequently sacred tales that have explanatory power regarding human experience. Still, we have a conscious and unconscious desire to rebuild or rediscover the lost perfect world.

Humans have a natural desire to live in an ordered world, and we miss that life. Early myths reflect a world that (hopefully) once was: a world of peace in which people lived without tension with one another and in which there was harmony and peace with both nature and the gods. When people separated from the gods, they lost perfection. Through their imaginations, they tried to restore belonging to the earth. Thus the original idea of the golden age was created.

It is widely believed that the classical Greek and Roman eras were the Golden Age of humanity, in which man, Earth, and nature lived together harmoniously. However, on further examination, it seems that the Greeks and Romans were the same as we are today: stubborn, unwilling to solve actual problems, and selfish. In Jared Diamond's book, *The Third Chimpanzee*, he writes, "It's now clear that pre-industrial societies have been exterminating species, destroying habitats and undermining their own existence for thousands of years" [2]. He then goes on to talk about how it is a false notion that it was the white, male, dominant cultures that civilized new territories. Rather, a more accurate representation is that they went to indigenous peoples' land, who many times lived in harmony with nature, and took them over, destroying that harmonious relationship with nature. He gives examples from when the British went to New Zealand and caused many animal extinctions to when the Portuguese found Madagascar and basically ruined it. According to Diamond, with innovation, there comes a certain point where things begin to worsen for a society in the long term.

His suggestion is that we need to look at the indigenous people of the lands we have colonized and see how they lived (and some of them still do live) in harmony with nature. We may live in the most advanced society in history, but are we the society that is the best off? At the end of the day, this is the simple question we should be concerned about, and we need to find good answers that will help humans survive on this planet.

2.2 Historical Examples: The Golden Age Again and Again

2.2.1 Greece

The Golden Age of *Greece* is described as the period lasting from approximately 500–300 BCE. Although it lasted for hardly more than a few centuries, it laid the foundations of western civilization.

Here is a list of Greek inventions and discoveries that have had deep impacts on Western culture and society [3]:

- Democracy
- The alphabet
- The library
- The Olympics
- Science and mathematics
- Architecture
- Mythology
- The lighthouse
- Standardized medicine
- Trial by jury
- The theater

Retrospectively and symbolically, the Greek Golden Age started with the surprising military success of the Greeks (mostly Athenians) against the huge Persian army in the Battle of Marathon. This era is also referred to as the "Age of Pericles," after the Athenian statesman who directed the affairs of Athens when she was at the height of her glory.

How should we characterize this period? While we cannot state that peace and stability were dominant, new forms of social organization emerged in a number of domains. It is very reasonable to consider them the basic ingredients of western civilization.

Government

Athens is rightfully credited as the birthplace of democracy. Democracy is defined as the "rule by the people," and rule by the majority is often implied. Earlier, in most societies, rule was forced by "strong men"—chiefs, emperors, kings, pharaohs, or warlords.

Well, Athens was the birthplace of democracy for the free males. Slavery was an accepted practice, and Aristotle himself stated that slavery is both natural and necessary. Later, however, the Stoic philosophers condemned slavery. In Athens, women did not have voting rights and could not own land. The natural place of woman was in the home, and their main mission was child rearing.

There are two distinct mechanisms for "becoming the boss": dominance and prestige. Dominance is an evolutionarily more ancient strategy based on the ability to intimidate other members in the group through *physical size* and *strength*. In dominance hierarchies, group members do not accept the social rank freely, only by coercion. Prestige, as a strategy, is evolutionarily younger, and is based on *skills* and *knowledge* as appraised by the community. Prestige hierarchies are maintained by the consent of the community, without pressure applied by particular members. Athens provides an important example in which prestige determined the leader of a community. Solon's economic reform reduced inequality, as he canceled all public and private debts, and introduced an income tax, taxing the rich more than he taxed the poor. Pericles made voting rights equal for all free and male citizens. (Given that women and non-free people were excluded, the percentage of the population that actually participated in the government was 10–20% of the total number of inhabitants.) He also adopted the practice of paying jurors so that common citizens were able to participate in the justice system.

It is very important to remember that Aristotle outlined the concept of *rule of law*, which prescribes that all people and institutions within a community are accountable to the same set of rules. The United States (and of course many other countries) adopted this rule. Another concept that came from the Aristotelian school is that of a *constitution*. A written constitution defines a common standard and prescribes the rules the people must follow.

Art

Pericles sought to make Athens beautiful, and he managed to convince the citizens' assembly to financially support artists and architects. Architecture, pottery, and sculpture advanced greatly during the Golden Age. The Parthenon, the temple of Athena Nike, and other buildings on the Acropolis became symbols of Athenian power and prosperity. Greek artists had an influence on modern art by emphasising the bodies of their subjects. The Discobolus of Myron is one of the most famous statues in the world. As we all know, the Nazis used this figure in their propaganda. We would hope that the association with Hitler has slowly disappeared.

Philosophy

Socrates broadened philosophy with logic and ethics. As he never "published" his ideas, his student, Plato, preserved the wisdom of Socrates in written accounts. Plato's treatise on government, *The Republic*, and his concept of ideal forms versus material reality are still intellectual sources of contemporary philosophical discussion. Aristotle's writings about physics and metaphysics (as well as politics and literature) have been strong motivations for scientists and philosophers for many centuries.

Drama and Literature

Drama was maybe the greatest intellectual invention of the ancient Greeks. It originated in religious ceremonies and became a complex art form. The playwrights grasped the fundamental aspects of life in both *tragedies* and *comedies*, which featured mythological and historical events. Aeschylus, who dramatized the story of Agamemnon, and Sophocles, author of the celebrated Oedipus tragedy, are still considered masters of dramaturgy. At least 44 plays from the ancient Greeks remain, and these serve as the foundation of Western theater.

Was It Really a Golden Age? If so, Why Did it End?

What brought the golden age to an end? At the end of the day, the answer is not surprising, even from a modern perspective: the combination of war and plague [4]. Is there anything we can learn from this historical example? Yes. First, we should try to predict and control plagues (and other natural disasters) as best as we can. The Athenians experienced a major drawback when a plague broke out in 430 BCE. About half of the Athenian population died, including Pericles himself. Second, while military theory suggests that war may have both positive and negative effects, the Peloponnesian War between Athens and Sparta certainly ended the prosperity of Athens and destroyed democracy. At first, there was an attempt to partially repair Athenian democracy: "A year after their defeat of Athens in 404 BC(E), the Spartans allowed the Athenians to replace the government of the Thirty Tyrants with a new democracy. The tyranny had been a terrible and bloody failure, and even the Spartans acknowledged that a moderate form of democracy would be preferable" [5].

The writings of the Greeks still inspire modern thinkers—examples from the Peloponnesian War come to mind, especially when it comes to (trying to) predict the future. Historians and political scientists started using the term "Thucydides Trap" a few years ago. The concept (explained by Graham Allison from Harvard's John F. Kennedy School of Government [6]) describes a situation where a rising rival is feared by a strong, reigning power, and the hegemon's fear of losing control leads to war. The Greek Thucydides (460–395 BC), also known as the father of scientific historiography, describes in his work on the Peloponnesian War how the growth of Athens frightened Sparta, then a superpower, leading to war between the two.

Since one of the purposes of this book is to help us "look into the past and predict the future," we should mention here the much discussed question of whether the United States and China can avoid the Thucydides Trap. Allison and his colleagues have identified 16 similar strategic dilemmas in modern history. Allison's concept, as it applies to the U.S.-China context, has been criticized by many. But the historian's important question is how to prevent two great powers from falling into the Thucydides trap by addressing the structural problems that led to the clash. We go back to this topic in Sect. 6.4.2.

History has its own meandering pathways. With its win in 338 BCE, a new power emerged, Macedonia. But this is a different story.

2.2.2 Rome

From the Greek Poets to Ovid

Over the course of centuries, Rome increased its territory and power, but the Golden Age of Rome—identified with the prosperity of the Roman Empire—ended around Marcus Aurelius's (121–180) death. Much of Roman art was candidly transferred from the Greeks. Visual artworks, such the marble statues and grand columns, originated fro m the traditions of the Hellenistic world. The Golden Age in Rome is reflected in the poems of Virgil and Ovid. Virgil's *Aeneid* commemorates the dignity of Rome's past. In addition to romantic poetry, historiography (like that produced by Livius and Tacitus) also flourished. Motivated by Hesiod's concepts of the historical periods, Ovid defined four Ages: The Gold, Bronze, Silver, and Iron Ages. It was the Iron age where all things went wrong.

Interplay: From the Golden Age to the Iron Age

Hesiod described in a poem, written in eighth century BCE, the continuous degeneration of the lives of men from a state of primitive innocence to evil. The Golden Age was timeless, and in the eternal spring, fruits grew without needing to plant seeds, so there was no agriculture. In the Silver Age, the eternal spring decomposed into three periods of Spring, Summer, and Winter. People did not live in Paradise anymore, and they had to seed and harvest. They had to build some shelter to keep themselves safe and warm, so men learned agriculture and architecture. In the Bronze Age, war appeared both as purpose and passion, but still people did not show impiety. In the Iron Age *all the things went wrong*. Selfishness and corruption emerged.

A Pessimistic Question: Do We Live in the Iron Age?

Ovid stated that in the Iron Age all the things went wrong. Modesty, truth and faith disappeared, and was replaced by fraud, violence, and lust for personal gain. Individuals did not have any true purpose. Family relations became fragile or worse. Mistrust between human beings became ubiquitous. The seeds for envy and hate and, consequently, war were planted. Ovid worried that there was no help for human kind anymore [7].

2.2.3   Judeo-Christianity

In the Judeo-Christian tradition, the Garden of Eden, also known as Terrestrial Paradise, as we all know, plays an important symbolic role in religious texts. Adam and Eve lived there prior to their expulsion, when the promise of eternal happiness came to an end. People, however, never fully accepted the eternal disappearance of the ideal universe, and cyclic struggle between "Paradise Lost" and "Paradise Regained" appears in many stories and myths. Most famously, the English poet John Milton (1608–1674) described the temptation of Adam and Eve by the fallen angel Satan and their expulsion from the Garden of Eden.

According to the creation myth in Genesis, which is part of both the traditions of Judaism and Christianity, the first man was put in the Garden of Eden. As we know, there are some competing interpretations regarding the exact location of the Garden Eden. According to the predominant Mesopotamian myth, it was located where the Tigris and Euphrates rivers run into the sea.

The story of the Garden of Eden is mostly about the beginning of human history, followed by the fall of man. It is also about the birth of the human knowledge and self-consciousness. The concept of the Garden of Eden has been reflected in the different ages of art history. One of oldest interpretations of the Garden of Eden is surprisingly geometrical and abstract and made in the Byzantine style. It is the blue ceiling mosaic of the Mausoleum of Galla Placidia in Ravenna. Circular motifs might represent flowers of the Garden of Eden.

Hieronymus Bosch's triptych *The Garden of Earthly Delights* should be read from left to right [8]. The left panel illustrates Paradise at the very moment in which God presents Eve to Adam. The large middle panel combines real and fantastical elements, while the rightmost is a nightmarish illustration of Hell.

The Garden of Eden with the Fall of Man was famously co-produced by Peter Paul Rubens and Jan Brueghel the Elder. Rubens painted the human figures and Breughel the animals, trees, and fruits. We see that while Adam sits under the Tree of Knowledge, Eve is grabbing the forbidden apple [9].

The continuation of the story is Masaccio's *The Expulsion from the Garden of Eden*, which grasps the moment when Adam and Eve leave the Garden of Eden in shame. Their emotions are well-reflected on their faces. We don't see any part of Eden in the background. Obviously, they are no longer part of the natural harmony of Paradise. The mythical Golden Age has been lost.

However, the story still fascinates artists, and the biblical Eden has captured the imaginations of many contemporary artists [10].

2.2.4 China

Historians like to identify a number of periods in the history of ancient *China* as golden ages. Famously, China's Tang Dynasty lasted from 618–907 CE. During this period, China experienced a time of peace and prosperity that made it one of the most powerful nations in the world. Technology and engineering advanced greatly. Perhaps the most important invention was woodblock printing, a new technology that made possible the mass production of books. Milestones were reached in the production of gunpowder and porcelain, and huge developments occurred in fields such as medicine and clockmaking.

It is difficult to imagine the worldview and cultural atmosphere of a time when poetry was a required field of study for those who wanted to enter the civil service. But this was the case during the period when Buddhism was popularized throughout China. Soon the collision between Buddhism and Confucianism started. Confucianism was declared to be the national religion, all other religions were banned, and many Buddhist monasteries were shut down.

Why did the Tang Dynasty decline and fail? The answer should be familiar to the modern Reader: government corruption and high taxes. People initiated rebellion against over-taxation, and the system never managed to fully recover.

It is interesting to know that China has about 50 million antique books. Repairing ancient books is tedious and demanding but rewarding work. In December 2019, a museum on ancient book repairing opened in Chengdu, the capital of the Sichuan province. We will return to this topic in Sect. 5.1.

2.2.5   Japan

The Heian period is called *Japan*'s Golden Age because during this time, aristocrats led a great flourishing of Japanese culture. The transfer of the capital from Nara to what was called Hian-kyo (present day Kyoto) was the starting point of this age. From 794, the new capital was a very well-designed city. It was laid out in a checkerboard pattern, and the aristocrats lived in beautiful mansions.

The centralized form of government was adopted from China, and the Fujiwara family controlled the imperial line and courts. As always in the golden ages, art flourished as new forms of painting and sculpture appeared. *Katakana* and *hiragana*, two Japanese syllabaries (similar to alphabets but representing syllables) emerged, and calligraphy was highly respected. Poetry was very important, even as a form of communication. Female writers had a leading role, particularly Lady Murasaki.

The Heian period, as any other historical era, eventually came to its end. Inequality between rich and poor dramatically increased. Civil war broke out, and the different regions of Japan fought for dominance. By 1185, the military elite started to dominate the country.

2.2.6   Greece Versus China: Some Comparisons

Research, notably by Jeremy Tanner, a British professor of Classical and Comparative Art, finds many striking parallels between the history of portrait painting in classical Greece and early imperial China [11]. The phenomenon is certainly interesting, since political organization is democratic in the first case and monarchical in the second. However, a striking parallel in the development of the two states is that the changes of a given historical period (such as increasing urbanisation or the peasantry becoming more independent and productive) unleashed new resources, including the emergence of a new cultural elite. Kleisthenes in Athens and Qin Shi Huangdi, the first emperor and the Han successors in China, channelled many of these changes into the creation of differentiated political institutions.

Tanner describes portraits as a form of reward symbolism. In this way, it is possible to understand how the emerging elite reinforces attachment to the newly forming state structures. Portraiture, as a political monument, has in many cases become a prestige-generating activity. In political terms, portraits aim to celebrate their subjects' achievements on behalf of the state, rewarding their participation in the life of the state. What is more, portrait painting has thus become a theater of social competition and a major medium of political integration and (of course) a focus of tension in both states.

The study of two traditions on equal terms always reflects changes in economic and political power, and changing intellectual trends. It takes account of similarities or differences without assuming one or the other to be superior. While comparisons between Greece and China do not always manage to leave out the narratives of the cultural superiority of the West, intellectual trends are changing [12].

2.3 Golden, but to Whom? the Dutch Controversy

In 2019 the Amsterdam Museum changed its policy regarding calling the seventeenth century of Netherlands' history the "Golden Age" [13]. Before this change was suggested, there was a general consensus in the Western world that not only Dutch art, but also the whole of the Netherlands, experienced its greatest prosperity during this period. The Dutch Republic "produced" such genius artists as Rembrandt and Vermeer, the philosopher Spinoza, and the scientist Van Leeuwenhoek, who invented microscopy. The economy was booming, and the Dutch East India Company, a trade organization, was the driving force of economic growth.

At the end of the nineteenth century, the term "Golden Age," as a label for the seventeenth century expressed the national pride of the relatively young Netherlands. Historians of that time found that "One would have to go back to the mists of antiquity, to the Athens of Pericles, to find a comparable example of such wide-ranging development in such a small tract of land" [13].

As we mentioned earlier with regard to the limiting of Athens' democracy to free men, recent similarly sharp debates have arisen regarding the Netherlands: A Golden Age to whom?

Tom van der Molen, the curator of the seventeenth century at the Museum, said: "The Golden Age occupies an important place in Western historiography that is strongly linked to national pride. But positive associations with the term such as prosperity, peace, opulence and innocence do not cover the charge of historical reality in this period. The term ignores the many negative sides of the seventeenth century such as poverty, war, forced labour and human trafficking."

What we see is nothing but the reevaluation of history. As it is often said, history is always written by the winners. We might agree that the seventeenth century in the Netherlands had nothing to do with Ovid's Golden Age. The country was involved in wars and initiated the colonization of others' lands. It might be true that "A Dutch Golden Age? That's Only Half the Story" [14]. One might argue that the now famous (or maybe infamous) "one percent" participated in the golden age, but how should we think about the remaining 99 percent? The term "seventeenth century" is more neutral, but we don't necessarily believe that we should throw out the baby with the bathwater. However, there is room to add new stories.

2.4 From the Golden Age to the Climate Crisis

The history of our deviation from the Golden Age is nothing more than the history of the climate crisis. Climate change is now with us, and the anthropogenic contribution to it is well-documented. While greenhouse gas emissions are a uniquely modern problem, it turns out that the narratives we use to understand the current moment have repetitive elements. The fear that any change to the planet might affect our survival is certainly not new.

As we know, myths suggest that humans and the Earth once coexisted in a perfect, timeless state. But humans that didn't live in the perfect Golden Age needed inventions to survive. Our advancements in technology have created a more significant gap between nature and humanity. We have factory farming, smog, blue light from our phones, and many other hazards. The positive story emphasizes progress, but the negative emphasizes the exploitation of the natural environment. It is just that now we have much more destructive tools in our hands. It is often said that we do not understand the destruction that we are causing until it is too late.

While our nostalgia for the rhythms of the natural world is timeless, the authors agree that eternal summer is not what we desire. (Even a vacation at the wonderful Adriatic seaside in Croatia might be suboptimal if it is too hot, as we experienced in the second week of August of 2021.) Our views resonate with those of Kathryn Wilson, a teacher and researcher of Classics: "As my students and I endured the extreme weather and discussed Virgil's depiction of farming, I realized how our thinking about the current climate crisis matches Virgil's about the loss of the Golden Age" [15].

The transition from the Golden Age to the Iron Age is the story of transformation of the timeless, static world view to the dynamic, irreversible perspective. In the Golden Age, humans did not experience any aging. The biological arrow of time is the product of the Iron Age.

Somehow we like to feel as the ancient Greek and Roman poets did: First there was harmony between humans and nature, and somehow the connection has been lost. We feel that we are in the same story …While technology has been the driving force to provide more food, clothes, and gadgets for our modern lives, the natural rhythms of life have disappeared. In our childhood, we never saw any grapes during the winter, but now there is no seasonal limit to our consumption. The price that we pay for the possibility "to buy anything at any time" is the loss of a natural harmony. Dear Reader: Accepting to eat "seasonally" and "locally" would provide less flexibility of choice but ensure more natural harmony.

The Dutch "Golden Age" and the Little Ice Age

Concern about the effects of climate change is certainly not new. The speed and global scale of current change is different from the previous eras, but variations in climate due to natural shifts previously occurred, even before human society established big industry. Paleoclimatologists have been able to uncover the permanent signs of past changes in temperature and precipitation. While these changes generally led to disasters, a young climate historian, Dagomar Degroot has suggested that the Dutch Republic, the precursor of present day Netherlands, managed to successfully react during the chilliest period of what is generally called the "Little Ice Age" [16, 17]. This climate period was caused by the combined effects of dust generated by huge volcanic eruptions and the reduced quantity of solar energy reaching the northern hemisphere. Sea ice expanded, sailing became more difficult, and sailors were forced to discover new pathways to substitute for the Northeast Passage and Northwest Passage. Sailors also found new wind patterns to speed up their ships, and of course, the length of their journeys became shorter.

While today we speak about the superiority of "local" foods and food consumption was very local before globalization, provinces of the Low Countries around the seventeenth century needed to import grain. When imports from the south decreased, grain commerce from the Baltic emerged, and grain import became a significant source of employment. Maritime transport was supplemented with transactions related to handling, storage, quality control, and manufacturing. Trade organizations became the driving force of economic development. Baltic grain was stockpiled in good years and was sold for healthy profits whenever food shortages plagued Europe. The legendary charities of the "Golden Age" provided a continuous supply of food for the poor people of the cities.

It is not surprising that the Dutch fought most of their wars on or around water. Their armies and fleets may have benefited even more from climatic cooling than their merchants. The Dutch flooded their own farmland to block Spanish and French invasions. Some of these floods would not have succeeded without the huge rains that were a consequence of the changed atmospheric flow.

As a result of the particular geographic location of the country, the Dutch learned that the environment can change and that people should find adaptive strategies for survival. The lesson that we can learn is that human adaptation to climate change may have positive effects. Still, we don't want to suggest that the proliferation of future orange groves in London might be a positive side effect of the climate crisis.

2.5 The Golden Age of Babies

It was so good once upon a time. What was it like then? Above all, it was safe. Everything was fine: It was warm, we had enough to eat, and we were cradled in the arms of our mothers. Ancestral (primal/primordial) trust is the belief that we can trust the world, that it is good to be in it, that it will hold us, and that we can rely on it [18]. This feeling and belief begins to take shape in the womb and later in a baby's life depends predominantly on a mother's reassuring and loving presence. If it is sufficiently developed over the first two years of life, a baby will begin to explore human relationships and the world with courage and confidence.

For a fetus, safety is the most important thing. If the mother is stressed or traumatized, a fetus can be psychologically damaged, which can later lead to mental illness. A newborn is totally dependent on its mother, and if it receives loving reactions and responses when it expresses its needs—for example, crying because of fear, pain, or hunger—then the baby will have a sense of being, a trust in existence, and a belief that it is loved and in a good place [19]. A mother who accepts her baby with unconditional love, rocking, and gentle caressing allows the infant to begin to develop a sense of self-love. This maternal response is more than mere care: It is a sensitive parental response to the specific needs of the baby. It is possible to provide for a child with tender but cool care, but this does not create an atmosphere around the baby in which he or she can address the world. This means that she or he cannot trust the world and, therefore, has much less trust in himself or herself. The child develops mistrust and fear. This lack of confidence leads to an identity crisis, which in most cases can be remedied later with therapy.

We have discussed whether we should include some comments on the allegedly Golden Age of babies around 5–7 months old. Zsuzsa's feeling is that this period is a Golden Age for parents, and not necessarily for the babies but for the parents. The belief in the minds of parents about the golden age of babies is certainly linked to the long-standing idealization of infancy, of childhood. Both beliefs were born in the minds of adults, and we have no means of exploring the minds of babies of a few months old. What we do know is that the Swedish educational writer Ellen Key, in her famous book, *The Century of the Child*, published in 1900, portrayed the coming centenary as the century of the child [20].

Although we cannot speak of a triumphant cult of the child as she described it, there is no doubt that the child plays an increasingly important role in the lives of middle-class families and small communities [21, 22]. Although the romantic myth of childlike innocence and purity has been reinforced, Dieter Lenzen (later the President of Universität Hamburg for many years) argued that the idealized world of the child is in fact an escape from reality for adults [23].

2.6 Messages from the Past

Why do we have the tendency to yearn for the past? Occasionally we think, "If only I could go back to that time, all of my problems would be solved and I would be much happier." The term *nostalgia* comes from Greek: "Nostos" means return and "algos" means pain. Odysseus is driven to return to his native Ithaca and to his Penelope. Jewish people from their Babylonian captivity lament the loss of their homeland: "Alongside Babylon's rivers we sat on the banks; we cried and cried, remembering the good old days in Zion" (Psalm 137:1). Nostalgia is often related to homesickness, helps connect us distant places and times, and forms a continuous narrative from seemingly isolated events. As we know from Marcel Proust "Remembrance of things past is not necessarily the remembrance of things as they were" [24].

Nostalgia towards a Golden Age is in the center of Woody Allen's 2011 movie *Midnight in Paris*. The main character, Gil Pender, a novelist, feels that the source of his unhappiness is that he lives now, and not in 1920s Paris. If only he could go back to that time, all of his problems would be solved and he would be much happier. His friend, Paul explains:

> Nostalgia is denial—denial of the painful present …the name for this denial is golden age thinking—the erroneous notion that a different time period is better than the one one's living in—it's a flaw in the romantic imagination of those people who find it difficult to cope with the present [25].

Gil has managed to time-travel back to 1920s Paris and meets Dali, Hemingway, Picasso, and other artistic celebrities. He also meets Adriana there. She does not see her age as the golden age of Paris, so they time-travel back to late-1800s Belle

Epoque, which seems to be a golden age between the horrible time of the Napoleonic wars and first World War. The story goes on, with the people of the Belle Epoque looking back to the Renaissance.

Nostalgia as a Driving Force Behind Remake and Repair

Two types of nostalgia, restorative nostalgia and reflective nostalgia, seem to have different functions [26, 27]. *Restorative* nostalgia refers to the type where you try to reconstruct or relive the way things were in the past. National and religious revivals belong to this class; the cult of confederate symbols may suggest at least an emotional return to the a supposed origin. Much of the Renaissance style can be labeled "restorative nostalgia" related to classical Greece and Rome.

Reflective nostalgia is based on the acceptance that the past is the past. I (P) recently had a positive experience. It happened that the Hungarian edition of *Ranking* got some media attention, and I got an email from Leslie, my former classmate from the elementary and middle schools and someone with whom I have not met for …60 years. I was somewhat envious of him as he was the obvious first in math in our class. He became a successful CEO of a big company and has had a full, happy life. Old boy networks are generally very dense, so it was an accident that somehow we never met after our youth. After Leslie wrote me, we organized a small group party a month later, and we played a little soccer (which we finished after the second downfall and before the first blood). We somehow managed to reconcile the distant past with our actual living conditions by vivifying our common stories from the pioneer camp and our actual world view. We recreated a piece of our past and became a little happier.

Reflective action may be even weaker. Maybe I have another classmate, who did not felt the necessity of making an active step to revive some elements of the past. Perhaps he just started to search an etiolated copy of the class journal we edited together many decades ago. I will never know.

Loneliness is a widespread problem both at psychological and societal levels. The pandemic-related lockdown appears to have triggered a wave of nostalgia. Old-fashioned board games became popular. Many of us watched the rebroadcasting of classic sporting events, concerts, and theater performances. Nostalgia helped many of us cope with the negative mood of loneliness by using some stabilizing mechanisms to be discussed later in Sect. 4.1.1. Initial data suggest that it is a cross-cultural phenomenon observed in China, the United Kingdom, and the United States [28].

2.7 Lessons Learned and Looking Forward

The central focus of this chapter was the age before things went wrong. The loss of Paradise as a result of expulsion from the Garden of Eden remains unprocessed in spite of the continuous reflections of artists in changing times. The Greco-Roman and the various Asian golden ages, and their falls, are a dominant part of human history.

It appears that people like to have the nostalgic feeling that it would have been better to live in some earlier period. Golden or not, history teaches us that there were better-than-average ages. While we already saw some explanations for how and why these periods ended, we are now ready to study the question: Why do things go wrong? As always in this book, object-level, person-level, and social-level phenomena will be discussed.

References

1. Colarusso CA (2011) The golden age of childhood: the elementary school years. Tue Nat Prod
2. Diamond JM (1992) The third chimpanzee: the evolution and future of the human animal. Harper Perennial, pp 317–338
3. 11 Greek Influences and Contributions to Today's Society (9 Apr 2018). Owlcation. https://owlcation.com/humanities/Greek-Influences-today
4. Hughes T (2020) Did the plague of Athens end the city's golden age? Battles of the ancients. https://turningpointsoftheancientworld.com/index.php/2020/04/05/did-the-plague-of-athens-end-the-citys-golden-age/
5. The Final End of Athenian Democracy (Accessed 17 Apr 2022). PBS.org. https://www.pbs.org/empires/thegreeks/background/48.html
6. Allison G (2019) Destined for war: can America and China escape Thucydides' trap? Scribe Publications
7. Four Ages of Man (by Ovid) (Accessed 17 Apr 2022). GreekGods.org https://www.greek-gods.org/mythology/four-ages-of-man.php
8. The Garden of Earthly Delights (Last updated 2021, November 20). Wikimedia Commons. https://commons.wikimedia.org/wiki/The_Garden_of_Earthly_Delights
9. Senguen E (2021) The Garden of Eden and Archaeological Evidence of Its Existence. Yoair Blog. https://www.yoair.com/blog/the-garden-of-eden-and-archaeological-evidence-of-its-existence/
10. Scanlan J (2014) Back to Eden: contemporary artists wander the garden. Museum of Biblical Art
11. Tanner J (2015) Portraits and politics in classical Greece and early imperial China: an institutional approach to comparative art. Art Hist 39(1):10–39. https://doi.org/10.1111/1467-8365.12191
12. Beercroft A (2016) Comparisons of Greece and China. Oxford Handbooks Online. https://doi.org/10.1093/oxfordhb/9780199935390.013.14
13. van der Molen T (2019) The problem of 'the Golden Age.' CODART. www.codart.nl/feature/curators-project/the-problem-of-the-golden-age/
14. Siegal N (2019) A dutch golden age? That's only half the story. The New York Times. https://www.nytimes.com/2019/10/25/arts/design/dutch-golden-age-and-colonialism.html
15. Wilson K (2020) The golden age and climate crisis. EIDOLON, Medium. https://eidolon.pub/the-golden-age-and-climate-crisis-e63aee7771b9
16. DeGroot D (2018) The frigid golden age: climate change, the little ice age, and the dutch Republic. Cambridge University Press, pp 1560–1720
17. DeGroot D (2018) A frigid golden age: can the society of rembrandt and vermeer teach us about global warming? Historical Climatology. https://www.historicalclimatology.com/features/a-frigid-golden-age-can-the-society-of-rembrandt-and-vermeer-teach-us-about-global-warming
18. Erikson E (1950) Childhood and society. Norton & Co, W.W
19. Winnicott D (1957) The child and the outside world. Tavistock
20. Key E (2000) The century of the child (1900). BibkioLife

21. Muller P (1973) Childhood's changing status over the centuries. In: Brockman LM, Whiteley JH, Zubak JP (eds) Child development: selected readings. McClelland and Stewart, pp 2–10
22. Ariès P (1960) L'Enfant et la vie familiale sous l'Ancien Règime. Éditions du Seuil
23. Lenzen D (1985) Mythologie der Kindheit. Rowolts Taschenbuch Verlag
24. Proust M (2003) 1913. In: Search of lost time, Modern Library, SLP edition
25. https://youtu.be/ERq0yzfGAM0
26. Boym S (2001) The future of Nostalgia. Basic Books
27. McDonald H (2016) The two faces of Nostalgia. Psychology Today. https://www.psychologytoday.com/us/blog/time-travelling-apollo/201606/the-two-faces-nostalgia
28. Zhou Z, Sedikides C, Mo T, Li W, Hong EK, Wildschut T (2021) The restorative power of Nostalgia: thwarting loneliness by raising happiness during the COVID-19 pandemic. *Social Psychological and Personality Science*. https://doi.org/10.1177/19485506211041830

Chapter 3
Why Do Things Go Wrong?

Abstract In this chapter, we discuss the mechanisms by which things go wrong. Our goal is to illustrate the omnipresence of this class of phenomena from objects to individual human lives to whole societies. We will start with the physics of irreversibility, followed by deterioration due to wear and tear. The next section is about a new phenomenon—i.e., the increasingly ubiquitous "burnout" occurring in workplaces. On a different scale, our review of the general mechanisms behind natural and societal disasters will be followed by a difficult subject: why and how very important personal relationships painfully decay so often. We finish the chapter with a short discussion of malignant polarization in large social groups.

3.1   The Meandering Pathways of Irreversibility

From Good to Bad

While this book is about the concept of "Repair," things can only be repaired if they were initially functioned.

In this chapter, we discuss the mechanisms by which things go wrong. Our goal is to illustrate the omnipresence of this class of phenomena from objects to individual human lives to whole societies. We will start with the physics of irreversibility, followed by deterioration due to wear and tear. The next section is about a new phenomenon—i.e., the increasingly ubiquitous "burnout" occurring in workplaces. On a different scale, our review of the general mechanisms behind natural and societal disasters will be followed by a difficult subject: why and how very important personal relationships painfully decay so often. We finish the chapter with a short discussion of malignant polarization in large social groups.

Natural macroscopic processes have directions. Soup cools down because the heat slowly dissipates into the environment. Milk turns sour. Fruits and vegetables rot. Warm emotions also cool. People age and die. Irreversible damage is relentlessly with us.

Linear concepts of time are based on the view that stories have a beginning, followed by some other events, and they have an end: past, present, and future. The ancient Persian Zarathustra (or Zoroaster) may be responsible for the appearance of the linear concept of time in Western thinking. Judaism adopted linear time concepts and created *historical thinking*: the idea that events can be ordered in sequence, from beginning to middle to end. Christianity and Islam inherited this view. (The cyclic time concept—the idea that the universe does not have a final state and exhibits periodic motion—appeared in many ancient cultures. Buddhism has the notion called "saṃsāra," the famed spinning of the Wheel of Life or the Birth-Death Cycle of Being).

While our experience suggests that spontaneous macroscopic processes are irreversible, the microscopic processes experienced by molecules are reversible, and Newtonian physics describes reversible and cyclic motions. In principle, you can visualize the reversibility of any process. Assume a film of the process has been made—by running the film backward through a movie projector, you see the reverse process, which is the same as running it forward. So we have two fields: Classical mechanics, which describes reversible motions, and the theory and practice of irreversibility, called thermodynamics.

The first and second laws of thermodynamics have fundamental importance. The first law states a principle known as conservation of energy. Different forms of energy can be converted into one other (say mechanical work to heat) but cannot be destroyed. The total energy of an isolated system is constant. The second law tells us that while the mechanical work can be converted into heat completely, the reverse transformation cannot be complete. Some energy always dissipates. "Heat can never pass from a cooler to a warmer body without some other change," as Rudolf Clausius (1822–1888) stated in 1854. Heat cannot be completely converted to work, and *entropy* measures this non-convertibility. The entropy of an isolated system has a tendency to increase, and it has a maximal value in a state of equilibrium. The second law of thermodynamics describes all natural spontaneous processes.

Thermodynamics is not just a theory of heat but a general systems theory of the physical world. In a somewhat broader context, the two fundamental laws of thermodynamics suggest constancy and change. The first law reflects nature's constancy; the second one assigns the direction of changes. C.P. Snow famously complained that "highly educated" humanists don't know anything about these laws:

A good many times I have been present at gatherings of people who, by the standards of the traditional culture, are thought highly educated and who have with considerable gusto been expressing their incredulity at the illiteracy of scientists. Once or twice I have been provoked and have asked the company how many of them could describe the Second Law of Thermodynamics. The response was cold: it was also negative. Yet I was asking something which is the scientific equivalent of: Have you read a work of Shakespeare's?

From C.P. Snow's *Two Cultures* [1]

Historically, the theoretical attempt to reconcile the irreversibility expressed by the second law of thermodynamics with thinking about reversible mechanical processes produced strong debates. The Viennese Ludwig Boltzmann (1844–1906) defined a quantity (expressed with molecular characteristics) with the property of never decreasing over time. This quantity is proportional to the macroscopically defined entropy (one of the most discussed yet abstract scientific concepts). Boltzmann used statistical assumptions to derive his equations. This was the period during which concepts like "probability" and "randomness" became legitimate notions in modern science.

Linear concepts of time are reflected in the idea of an *arrow of time*. Arthur Stanley Eddington (1882–1944), a very influential British physicist and philosopher, coined the term. The irreversibility of macroscopic processes is manifested in the *thermodynamic arrow of time*. An increase in entropy characterizes all processes that transition to disordered states. The heat death of the universe is a cosmological possibility. Entropy is a measure of disorder, and spontaneous isolated systems irreversibly increase entropy, as the famous second law of thermodynamics expresses. Disorder may be reduced in non-isolated systems. (Of course, the total entropy, that of the open system and of the environment, would not decrease). Energy flowing through the system makes it possible to produce "dissipative structures" in an open system, which are not possible in isolated systems. Living systems are *not* in equilibrium with their environment. Our temperature is fortunately not the same as room temperature, so we are in a living state. Still, we will die, and the theory of evolution gives an explanation of why.

Aging and death from evolutionary perspective

Fabian and Flatt provide an evolutionary perspective on aging and death, explaining two specific hypotheses: "Today, it is clear that aging is not a positively selected, programmed death process, and has not evolved for 'the good of the species.' Instead, aging is a feature of life that exists because selection is weak and ineffective at maintaining survival, reproduction, and somatic repair at old age. Based on the observation that the force of selection declines as a function of age, two main hypotheses have been formulated to explain why organisms grow old and die: the mutation accumulation (MA) and the antagonistic pleiotropy (AP) hypotheses. Under MA, aging evolves because selection cannot efficiently eliminate deleterious mutations that manifest themselves only late in life. Under AP, aging evolves as a maladaptive byproduct of selection for increased fitness early in life, with the beneficial early-life effects being genetically coupled to deleterious late-life effects that cause aging. Aging clearly shortens lifespan, but lifespan is also shaped by selection for an increased number of lifetime reproductive events. The evolution of lifespan is therefore a balance between selective factors that extend the reproductive period and components of intrinsic mortality that shorten it. Whether there exist truly immortal organisms is controversial, and recent evidence suggests in fact that aging might be an inevitable property of all cellular life" [2].

The cosmological arrow of time describes the direction of the expansion of the universe, while the psychological arrow of time expresses the feeling that we travel from the past to the future as we grow old. An isolated system that does not have any interaction with its environment implies decay, and this fact led to the hypothesis of the *heat death of the universe*. In general, irreversibility does not necessarily imply decay. Evolution is also irreversible. The pseudo-contradiction between entropy-increasing thermodynamics and complexity-increasing evolution is not real, since entropy increases in closed system while complexity increases in open ones. Two major narratives that arose during the Victorian period (those proposed by Charles Darwin (1809–1882) and Karl Marx (1818–1883)) stated (for better or worse) the existence of the biological and historical arrows of time.

However, in this chapter we focus on the mechanisms by which things get worse. *Hints for the Overworked*, which focused on mental and emotional wear and tear and was first published in 1871, was a best-seller written by the physician Weir Mitchell on the etiology and treatment of neurasthenia [3]. In the next section, we discuss *wear and tear* as a general mechanism.

3.2 Wear and Tear

Wear and tear is the damage to an object that gradually and inevitably occurs as a result of normal use. It is different from an abrupt break. We will review wear and tear mechanisms and their effects somewhat arbitrarily in three very different areas: objects, with spontaneous glass breakage; people, with overwork and burn out; and societies, with their decline and collapse.

3.2.1 Spontaneous Glass Breakage

Spontaneous glass breakage is a phenomenon by which glass may break without any apparent reason. Several highly publicized incidents of spontaneous glass breaking and falling from high-rise buildings have been reported in cities including Austin, Chicago, Las Vegas, and Toronto, among others.

An undergraduate research assistant of mine (P), Jojo Lee (from Auckland, New Zealand), who helped me with reviewing literature during the pandemic, found a relevant story published recently in the Otego Daily Times [4]. A retired couple had just finished their breakfast when they heard a frightening sound from their kitchen. The toughened glass lid of their decades-old electric frying pan spontaneously exploded. Many common household items were once made from toughened glass, and warnings now are attached to products made from these types of glasses.

Window and balcony glass spontaneously breaks and falls from tall buildings. The phenomenon affects tempered glass, where the center of the glass is under extreme tension. A variety of tiny, hard-to-notice factors lead to the breakage [5]:

- Damage to the edges of the glass can make it vulnerable
- Stressors from temperature changes to wind to building movement, as well as other factors, lead the glass to expand and contract; such movement can put the glass under unsustainable stress, leading to breakage
- Badly fitted metal frames can apply extra stress to the glass where it contacts the frame
- Impurities in the glass can change in mass relative to the glass itself, applying further stress.

All of these factors together can result in breakage that appears spontaneous but was built up over time.

Such breakage supports the idea that things go wrong when *destructive forces* are not balanced by *healing forces*. The glass breaks because of *wear and tear*— many stress factors accumulate over time, leading to macroscopic, bad outcomes. The stress factors themselves are not always visible.

To avoid catastrophes like spontaneous glass breakage, where the precursors of the event are difficult to see, we can try to eliminate vulnerabilities. We can do a better job of identifying vulnerable objects and dispose of them before they pose a danger. This is what some specific heating processes attempt to do.

If the potential catastrophe is as dangerous as spontaneous glass breakage, we should prepare to replace fragile objects before they break! Now we will take a huge leap and turn from fragile glass to fragile humans.

3.2.2 Overwork and Burn Out

Wear and tear is a normal part of human life. However, its rather extreme form is what we call *burn out*.

A highly circulated article recently analyzed the question of why millennials have become the *burn-out generation* [6]. As usual, the coin has (at least) two sides: They are often stereotyped as lazy and entitled, but they are also said to be workaholics. Young people, especially women, are becoming increasingly career-focused. Highly skilled careers are very competitive, and young people expect to change jobs multiple times during their lives. But at the same time, some young people are experiencing a disconnect between their career expectations and the true realities of the workforce. These are new, important changes compared to the experiences of the previous generation. Many ambitious young people are sacrificing their social lives, hobbies, and even their physical and mental health to get a leg up in the career they want. Still, financially speaking, most of them lag far behind where their parents were when they were the same age. They have far smaller savings, far less equity, far less stability, and far, far more student debt.

People feel burnt out when their internal resources have been exhausted, and since reduced professional activity can be a consequence, it is difficult to stop what becomes a vicious cycle. As the Korean-German philosopher Byung-Chul Han discussed in his bestselling book, the big thing is that the new burn outs are exploited by themselves and not by their supervisors [7]. People forget about the saturation effect: It is not possible to increase one's investment of time without any limit. One's super-activity increases through self-amplifying, positive feedback without adopting any self-stabilizing, negative feedback mechanism (say, relaxation). However, it is not a good strategy for success. You cannot work 100 hours per week without paying the price.

Science is based on quantification and measurements, so it is a natural question to ask how to measure burn out. The Maslach Burnout Inventory (MBI) was created to provide an answer [8]. It takes into account three criteria: emotional exhaustion, depersonalization, and personal accomplishment. The first can be identified with total lack of energy, the second with some feelings of cynicism or negativity toward a job, and the third with reduced efficacy or success at work. Participants get scores in all three areas along a continuum, from more positive to more negative. A scientifically defined burn out profile requires a negative score in *all three* subcategories. Real burn out cannot be repaired with a wellness retreat or vacation. It often leads (for better or worse!) to career change. Data suggest that about 10% of the employees in all professions experience burn out. The pandemic raised this number without a doubt, and although we don't have firm data, experts believe the percentage might now be close to 20.

Burn out now looks like a global problem. It has been the subject of research and policy responses across Europe [9]. The COVID-19 pandemic—along with major political disruptions and natural disasters ranging from hurricanes to wildfires—has led to soaring rates of potential burn out. One (not necessarily statistically representative) 2020 study found 76% of U.S. employees were experiencing work-related burn out [10]. (Please note: Numbers can be dangerous!)

In East Asian countries, the increase in "overwork deaths" is an alarming signal. Heart attacks and mental stress are often involved. Japan has some of the longest working hours in the world, and it is well known that some young Japanese workers are literally working themselves to death. There is a special expression in Japan for overwork-related death: *karoshi*. *Premium Friday* was a government-driven campaign to end work at 3 pm on the last Friday of each month to help workers cope. It was a recommendation only. In addition, as the COVID-19 situation required people to stay at home and not travel, rather than go out and have fun, Premium Friday has fallen out of fashion.

There is a viral, somewhat ironic slogan among Chinese computer programmers that propagated on the software repository GitHub: "996.ICU." It encodes the idea, "Work 9am to 9pm, six days per week—end up sick in an ICU bed somewhere." ICU here is of course a reference to the Intensive Care Unit, and the message is that by following the 996 work schedule, you are risking putting yourself into the ICU. While labor laws in China do not permit people to work more than eight hours per day and 44 hours per week, this law is very often violated. Regional surveys suggest that more than half of respondents work overtime every single day.

Measurement in psychology is not a trivial procedure, so it is understandable that there are a number of different tools to assess burn out. MBI itself has variations in China, Japan, and other contexts. A reliable measurement does not solve the problem, but at least it makes it more visible. If many cases of burn out are identified in an organization, the problem is not with the individuals—they just react to the environment! Workplaces with unmanageable workloads can be identified by measuring MBI. People may simply be working more hours because they are worried they will not get a promotion or will even lose their job.

Human burn out seems to arise when negative forces are not balanced out by positive ones. There are patterns of rapid drops in performance and engagement as a result of cumulative stressors. There is a recurring pattern of not noticing that a person needs rest until things go catastrophically wrong.

Here we are. Repair after-the-fact does not seem nearly as effective as preventing deterioration in the first place. We will discuss some recovery strategies in Section 4.1.

3.2.3 The Road to Societal Collapse

Jared Diamond, a deservedly famous geography professor at the University of California Los Angeles, raised a question by analyzing the end days of disappearing civilizations (for example, the Maya): Why did they die out, and what lessons can we learn to understand their vulnerability? According to popular theory, the Easter Islands extinction could presumably have been prevented if the natives would have made boats out of the youngest—instead of the oldest—trees. More precisely, they themselves caused an ecological and civilizational catastrophe that led to their extinction. It is understandable that fishing requires boats, which can be built of wood. We ask cautiously, however: Is it possible that not one of the islanders saw that the disappearance of the trees would create a new problem—that they would not be able to produce the equipment necessary to obtain their most important food? They switched from fishing to farming, but it was not well-regulated, so the soil was exploited and their civilization was destroyed.

Generational Differences

There might be a difference in how boomers and younger generations perceive the chances of global societal collapse. Boomers either experienced or lived in the aftermath of world wars. Their ideas of civilizational and existential crises are linked to the possible outbreak of nuclear war. A few world leaders gained, for the first time in history, the ability to kill hundreds of millions of people. This was an alarming signal in a robust trend: As technology improves and the world economy grows, it becomes easier to cause destruction on an ever-larger scale. This possibility appeared in pop culture: *Fail Safe* was a famed thriller in the early 1960s that presented a story in which a technological glitch accidentally caused a thermonuclear first strike.

As we are learning from younger generations, their ideas of strain and collapse are related to different things: environmental destruction, climate change, surveillance technology, pandemics, extractive economies, and, perhaps in the future, technologies like transformative artificial intelligence (AI) and bioweapons. (Many of us predict that transformative AI might be as significant as the agricultural or industrial revolutions were.) While the probability of the COVID-19 lab-leak hypothesis is not very high, technology has granted the possibility of creating bioengineered viruses [11]. It is important to bring awareness of new dangers to new generations, which are so used to looking into the future: New transformative technologies may provide a radically better future, but they may, at the same time, increase the probability of catastrophic risks.

Collapsology

There is an old-new scientific field called collapsology. It adopts the perspective of "looking into the past, predicting the future." Historical analyses of societal catastrophes might help us foresee the precursors to the potential collapse of industrialized global society. There are lessons learned about the causes (both internal and external) of societal collapse. Here is a somewhat arbitrary list of causes:

- Resource depletion
- Climate change
- Reduced support from friendly neighboring societies
- Increased pressure from hostile neighboring societies
- Political, economic, social, and cultural factors that dictate whether the society is likely to perceive and fix its problems.

Societal collapse tends to happen rapidly. Rather than winding down slowly after their peak, societies fall suddenly and precipitously. This pattern is likely to occur when there is a mismatch between resource availability and resource consumption or economic outlays and economic potential.

This illuminates a key point: Human societies are *problem-solving* organizations. If there is a conflict between the interests of the decision-making elites in the short run and the long-term interests of society as a whole, problem solving fails. More precisely, when a society as a whole does not have the cognitive capacity to provide an appropriate response in a critical situations, collapse might happen. The collapse of the Soviet Union is a textbook example of failure due to a lack of problem-solving ability.

> An old Jewish joke told in the 1960s
>
> The Lord is unhappy with the sins of the world and determined to destroy humanity with a flood in a week. He also communicates this to the peoples of the world in a thunderous celestial voice. This is where the world's leading politicians turn to their peoples. The U.S. president gives a televised speech: "My American brothers! We have a week to try to get to heaven through fasting, penance, and charity!"
>
> The Soviet party secretary also speaks:
> "Comrades! Open the warehouses, take the food and the vodka home! We have a week to try to live well!"
>
> The Israeli prime minister's speech is brief:
> "Jews! We have a week to learn to live underwater!"

It is true that the joke does not tell whether the learning experiment was successful. But the lesson is that the only chance of survival comes with facing reality and trying to find an appropriate strategy to cope with it.

3.3 Extreme Events and Predictability

3.3.1 Extreme Events

How Are Extreme Events Generated?

Extreme events, both in nature and society—such as earthquakes, landslides, wildfires, stock market crashes, the destruction of very tall buildings, engineering failures, and epidemics—may appear to be surprising phenomena whose occurrences do not follow any rules. Of course, these kinds of extreme events are rare, but they dramatically influence our everyday lives. Can we understand, assess, predict, and control these events? *Complex systems theory* offers a perspective that aims to understand the mechanisms behind emerging patterns. As a consequence of natural and social crises, the occurrences of rare, large, and extreme events are now the focus of extensive mathematical analysis.

We should realize that the adjective "extreme" was originally semantically neutral. It might label deviation in a positive direction (say, extremely productive or successful people). It is common to hear questions such as "What is the probability of having a big earthquake in Iceland within a year?" or "How large might a possible stock market crash be tomorrow?" Or to use a positive example: "What are the odds of winning the lottery?" (The early history of probability theory is strongly motivated by the desire to win in gambling.) The study of earthquake eruptions, the onset of epileptic seizures, and stock market crashes has traditionally been investigated by very different disciplines that diverge sharply in their scientific cultures. The complex systems approach emphasizes the similarities among these events and offers some common methods for predicting the behavior of such systems and understanding the inherent limits of their predictability.

In order to control and manage extreme events, we should first understand the mechanisms that generate these phenomena. One possibility is to say that big earthquakes are nothing but small earthquakes that do not stop. The consequence would be that these critical events would inherently be unpredictable, since they would not have any precursors. Another scenario is that catastrophic events, or at least a class of them, result from accumulating or amplifying cascades. Were this hypothesis true, predicting these events might be possible. (Unpredictability is related to "self-organized criticality" and some predictability to "intermittent criticality" [12, 13].)

Uncompensated Positive Feedback

Feedback is a process whereby some proportion of the output signal of a system is passed (fed back) to the input. So, the system itself contains a loop. Feedback mechanisms fundamentally influence the dynamic behavior of a system. The terms "negative" and "positive" feedback loops derive from control theory and describe how a system's behavior produces or diverges from equilibrium.

Roughly speaking, negative feedback reduces the error or deviation from a goal state and therefore has stabilizing effects. It ensures the system functions close to what is expected (for example, a thermostat ensures that the temperature of a room shows only minor deviations around the desired value by shutting off heat when the room gets too warm.) Positive feedback, which increases the deviation from an initial state, has destabilizing effects.

Natural, technological, and social systems are full of feedback mechanisms. Specifically, in an economy, there are many feedback loops that drive the system toward an equilibrium of supply and demand. Positive feedback loops, such as those caused by our tendency to imitate each other's behavior, can lead to explosive growth in prices, followed by an inevitable bursting of the bubble, which we discuss below. Uncompensated positive feedback seems to be a general mechanism behind phenomena like earthquakes, stock market crashes, hyperinflation, epileptic seizures, and political instabilities.

3.3.2 Too Much Growth Is Just Too Much

Unbounded Growth

Growth processes describe the increase of different physical, chemical, biological, economical, or other state variables like size, chemical concentration, population density, and price. Figure 3.1 shows three different growth patterns.

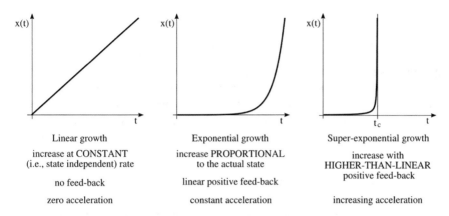

Fig. 3.1 Unbounded growth: finite and infinite time singularities

The first plot describes *linear growth*. Growth happens in the same amount in each time step. An example is income that grows with each deposit from an external source (some college students can identify this source as their dads). The velocity of the increase does not depend on the current quantity. The second plot reflects exponential growth. During the pandemic, "exponential" became a household term, and everybody learned how exponential growth works. The increase is greater with passing time, and the greater the current quantity, the larger the increase. More specifically, the increase is proportional to what you have. It implements a linear positive feedback between the velocity and the quantity itself. The implication of the feedback is that the process is accelerating. The feedback has one more consequence: The acceleration (the change in velocity) is constant. What is the common feature of the linear and exponential growth? When time tends toward infinity, growth also tends toward infinity.

What is more rapid than exponential growth? The super-exponential, as the third plot illustrates. Characteristically, the "doubling time"—the time necessary to double the quantity—tends to zero after a finite period. In the third growth pattern, the positive feedback is "higher-than-linear," which seems to be a general mechanism of the phenomenon known as *finite time singularities*. The absence of stabilizing effects attributable to negative feedback mechanisms may lead to catastrophic consequences like *explosion*. The technical term finite time singularity, roughly speaking, means that a dynamic variable achieves an infinite value during finite time. This process is qualitatively different from the exponential growth, since in the latter, infinite values can only be attained during infinite time.

Economic Growth: Should We or Should We Not?

Let's start with the extreme statement "growth is a moral imperative." In principle, this seems to be the key to making everybody happy, healthy, and rich. A very influential representative of this vision is the economist Tyler Cowen [14]. Neo-classical economic theory has assumed that the economy is capable of generating infinite wealth because market forces and technological innovation can be combined to guarantee innovation and more efficient production. But what happens if crude materials and, more generally, natural resources deplete? Concerns like these have generated the view on the other side of the spectrum. Here we find different versions of the *degrowth movement*. According to this perspective, growth is an unhealthy obsession. Kate Raworth, a British economist, presents the *doughnut* as a new economic model of limited growth [15]. The inner ring of the doughnut is defined or calculated as the minimum we need to lead a good life. The United Nations General Assembly set a series of Sustainable Development Goals in 2015, which in part help define "the good life." They specify the availability of food and clean water and minimal levels of housing, sanitation, energy, education, healthcare, gender equality, income, and political representation. The outer edge of the doughnut represents the ecological ceiling specified by a sustainable environment. It is determined by some acceptable level of damage to the climate, soils, oceans, the ozone layer, freshwater, and biodiversity. Of course, the doughnut is a *normative* model: It does not describe how the world works but rather how it should work. The model suggests that we should stay between the inner and outer edges of the doughnut—inside the doughnut. The doughnut model is being tested in practice in Amsterdam to improve the post-COVID-19 economy (see, for example, [16]).

We, the authors, still do not know better options for compromising between the two views described above or how to define what the title of the now-50-year-old famous book *The Limits to Growth* suggested [17]. Based on computer simulations of the interactions among five variables (population, food production, industrialization, pollution, and consumption of nonrenewable natural resources), the authors of *The Limits to Growth* warned that, by continuing current patterns, sudden and uncontrollable decline could happen to both population and industrial capacity in just a few generations. While the model was somewhat oversimplified, and could not grasp the quantitative details, its basic message is still valid: The "business as usual" model leads us to the edge of collapse. Growth that is too rapid is not stable. People may become euphoric as they see super-exponential growth. However, rapid increase is the precursor to crash. By the way, speaking about crash it came to our mind:

One more joke from behind the iron curtain: Where is capitalism? On the edge of the abyss. And socialism? One step ahead.

From Finite-time Singularity to Crash

The absence of negative feedback mechanisms may generate catastrophe. Extreme events like mechanical ruptures, earthquakes, stock market crashes, hyperinflation, virus propagation, and epileptic seizures are related to unbalanced super-exponential growth [18].

Conventional economic theory, which is based on the equilibrium between supply and demand, holds that decreasing demand is compensated by increasing supply. The complex systems approach, however, suggests that, partially due to our susceptibility to imitate each others behavior, there may be a period during which both demand and supply increase, which leads to singularities. Equilibrium theory works well when negative feedback expresses its stabilizing effect on the increase caused by positive feedback. While in "normal situations" the activities of "buyers" and "sellers" neutralize each other, in "critical situations" there is a cooperative effect due to imitative behavior ("everybody wants to buy it since everybody else has already bought it"). So the positive feedback is higher-than-linear. Super-exponential increase (due to irrational expectation) cannot be continued for "ever" due to the unstable nature of this process, and the increase is unsustainable. Consequently, it should be followed by a compensatory process (i.e., stock market crash).

The Tulip Bulb Mania

What is generally labeled the first economic bubble emerged during Holland's Golden Age, around 1635. Tulip bulbs were first bought for their aesthetic value, but as their prices increased, they became the subject of buying and selling for the purposes of making (large) profits.

There was a month during which the value of tulips increased twenty-fold. At a certain point, the Dutch government attempted to control the mania. After its regulatory actions, some informed speculator realized that the price could not become further inflated and started to sell bulbs. Other people soon noticed that the demand for tulips could not be sustained. Their attitude propagated very rapidly among people interested in business, and soon panic, a collective social phenomenon, emerged. During a period of six weeks there was a 90% reduction in the price of tulips, see Fig. 3.2.

Uncompensated Positive Feedback: Epilepsy and Financial Crisis

Two types of financial crises, namely stock market crashes and hyperinflation, are onset by positive feedback between the actual and the expected growth rate [13]. Large stock market crashes are social analogs of big earthquakes and of the onset of epileptic seizures (as with all analogies, this also has an appropriate scope and limits). Didier Sornette's view is that stock market crashes are not induced by single, local events (such as a raise in interest values or other governmental regulations) but are due to the unsustainable velocity of price increases. These speculative increases will make the system more and more unstable. Finally, the market collapses with the introduction of any small disturbance. Unsustainable velocity of price increases resembles the over-excitation of the epileptic brain. Indeed, the mechanism of vicious cycles characterizes both epilepsy and financial crises.

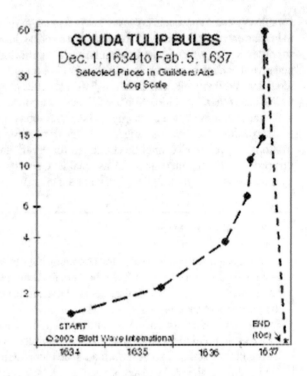

Fig. 3.2 Price dynamics of the rise and fall of the Tulip Bulb Mania. Adapted from http://www.stock-market-crash.net/tulip-mania.htm. ©Elliot Wave International

A Few Words About the Neurobiology of Epileptogenesis

The expression "Seizures Beget Seizures," for explaining the onset and progression of epileptic seizures, goes back to Sir William Gowers (1845 – 1915), a nineteenth-century British neurologist. Using modern terminology, seizures are involved in positive feedback loops. Complex systems research puts epilepsy into a broader perspective: As a complex systems disorder, epilepsy poses challenges for identifying single mechanisms or processes that are required for epileptogenesis or progression. An attempt to identify a list of specific molecular or cellular mechanisms responsible for the disease's progression is likely difficult because emergent phenomena such as network synchronization are not likely to be universally or linearly related to the specific defect [19, 20].

The stable dynamic operation of the brain is based on the balance of excitatory and inhibitory interactions. The impairment of inhibitory synaptic transmission relates to the onset of epileptic seizures. Epileptic activity occurs in a population of neurons when the membrane potentials of the neurons are "abnormally" synchronized. Both experiments and theoretical studies suggest the existence of a general synchronization mechanism in the hippocampal CA3 region. Synaptic inhibition regulates the firing of pyramidal neurons. In experimental situations, inhibition may be reduced by applying drugs to block (mostly) GABA-A receptors. (GABA is the most important inhibitory transmitter.) If inhibition falls below a critical level, the degree of synchrony exceeds the threshold of normal patterns, and the system's dynamics switch to an epileptic pattern. Collective phenomena occurring in neural networks, such as the case of disinhibition-induced epilepsy, have been studied successfully by combining physiological and computational methods.

How can economics exploit the analogy? There are at least two fields to which developed methods can be transferred: detection of precursors and new therapeutic strategies. Extreme events theory offers a method for the early prediction of epileptic seizures by studying the statistical properties of electroencephalographic (EEG) recordings [21]. It is useful to know that similar statistical properties (i.e., large price fluctuations during a short period) also relate to financial crises.

Can we widen the limits of our ability to predict extreme events like earthquakes, financial crashes, and epileptic seizures? One general observation is that extreme events often occur as a consequence of the impairment of the control system. The theory of complex systems suggests that extreme events may be predicted by detecting their precursors and that there are methodological similarities for analyzing and modeling different "critical events" that occur in physical, biological, and social systems. There are promising initial results and many open problems. The title of a recent paper expresses well the dilemma facing many theoreticians: "Can we predict the unpredictable?" [22].

What could be more unpredictable than human relationships? Next we will discuss these issues through the lens of social psychology and whether we can see the precursors of damage to relationships. What are our chances to "save" a relationship, and when should we decide that it is better let it go?

3.4 Patterns of Damaged Relationships

Shakespeare's Sonnets, 117:

Accuse me thus: that I have scanted all
Wherein I should your great deserts repay,
Forgot upon your dearest love to call,
Whereto all bonds do tie me day by day;
That I have frequent been with unknown minds,
And given to time your own dear-purchased right;
That I have hoisted sail to all the winds
Which should transport me farthest from your sight.
Book both my willfulness and errors down,
And on just proof surmise accumulate;
Bring me within the level of your frown,
But shoot not at me in your wakened hate,
Since my appeal says I did strive to prove
The constancy and virtue of your love. [22]

3.4.1 Maintaining Stable Relationships

It is difficult to precisely define relationships, partly because many of them overlap: A colleague can be a friend, a spouse a business partner, a boss a lover. Intimacy is a key feature of most close relationships (romantic, platonic, or otherwise), but think of it not primarily as a positive feeling but as a state of reciprocity of varying intensity and frequency. The attitudes of those in a relationship—and the degree to which such attitudes are reciprocated—determine how much their thoughts, communication, emotions, behavior, and actions influence each other and each other's decisions and plans. Above all, these attitudes and reciprocity determine their desire to maintain that relationship.

One thing is certain: We need stable relationships. It is best for our mental health since separation has high emotional, social, and physical costs. But, of course, even long-lasting relationships have moments of crisis. We should realize that fading romantic love is not a sufficient reason to end a relationship. Making long-term commitments is healthy.

Meryl Streep and Tommy Lee Jones play a couple in *Hope Springs* (2012), a film about reviving a troubled relationship. Their therapist, Dr. Feld, encourages them that, even in the best of marriages, there are years worse than others. However, he suggests that they not give up in the fight to preserve their marriage.

Disturbances always happen, and in Sect. 4.1.1 of the next chapter, we will discuss the general concept of stability. As concerns stabilizing initially, slightly troubled relationships, several questions must be answered.

What has gone wrong? How have we gotten this far? Is it that we expected something different from the relationship initially or our expectations changed over time? What are the signs when a connection is compromised? We may have been wearing a blindfold, not noticing that there was a problem with the relationship, that gaps were accumulating. As we have seen in Sect. 3.3.1, minor differences can be dramatically amplified if there is mutual positive feedback without any stabilizing intervention.

In his new comprehensive book about friendships, Robin Dunbar recalls an extensive series of experiments carried out by British social psychologist Michael Argyle (1925 − 2002) and his coworkers in the 1980s about the rules of friendship [23, 24]. These experiments helped identify six key rules that are essential for maintaining stable relationships:

1. Standing up for the friend in their absence
2. Sharing important news with the friend
3. Providing emotional support when it is needed
4. Trusting and confiding in each other
5. Volunteering help when it is required
6. Making an effort to make the other person happy.

Argyle and his fellow researchers found that the more of these six rules that partners break, the more they weaken the friendship. They also found that, when recalling relationships that had ended, respondents tended to attribute negative behavior to the other person and positive behavior to themselves. This is typical behavior according to the psychological literature. (More specifically, this refers to a concept called *fundamental attribution error*. Fundamental attribution error occurs when people underestimate the specific conditions in which they experience another individual's behavior and overemphasize that individual's personality. So, for example, somebody may block you on the road since she is running to a hospital to see a loved one and not because she had bad intentions. We all know this reasoning, as Dunbar writes: "It can't be me that is wrong, so it must be you.").

3.4.2 Why Do Relationships Break Down?

The most important gift of a relationship is psychological closeness. By opening up, we get to know each other better, which improves our cooperation, our ability to meet each other's needs, and increases trust. In romantic relationships, it is emphatically true that intimacy brings people together, and commitment based on feelings and actions holds them together. The longer the romantic relationship and the more committed it is, the greater the "investment" (shared property, shared friends, etc.) and the greater the 'cost" of breaking up.

In their survey, Dunbar and his fellow researchers offered respondents 11 reasons why a relationship might break down:

- Lack of caring
- Poor communication
- Drifted apart
- Jealousy
- Problems with alcohol or drugs
- Anxiety about the relationship
- Competition from rivals
- "Stirring" by other people
- Tiredness
- Misunderstandings
- Cultural differences between the couple.

The three most frequent responses were lack of caring, poor communication, and jealousy.

Underlying the first two reasons are three other fundamental and interconnected reasons for the breakdown of relationships. The first is the reduction in the time spent in the relationship: One member feels that the relationship is secure and that there is no need for the frequent meetings, conversations, and interactions that used to take place. This relates to a second reason: a lack of attention. If this becomes persistent, there is a good chance that, third, misunderstandings (also on Dunbar's list) will increase. And misunderstandings can lead to jealousy even in very small ways.

Both the theory of transitions (as we will explain in the Chap. 5) and data show that relationships break down either gradually or by some abrupt explosion. Some warning signals of the fading of a relationship will be mentioned in Subsect. 3.4.3. But first a little about Zsuzsa's experience.

3.4.2.1   Some Very Short Stories

Francesca, Zsuzsa's sociology student, conducted in-depth interviews with around two dozen people of different ages, genders, and professions. Her analysis showed that there are some typical patterns in relationship breakdowns.

A middle-aged woman named Eve mentioned a *lack of balance*. Eve gave up her professional career early in her marriage to care for her children and the household, and she devoted her time to looking after the family. In contrast, her husband looked after the finances. The children were in their teens when it became clear that the years had created a vast emotional distance between her and her husband, seemingly unnoticed over the years.

Helen and Kate were old friends, and in their early twenties, they had joined a left-wing student union at the same time. However, Kate's *political attitudes changed* over two or three years, and their conversations first turned into offensive debates and arguments and then resulted in a break-up.

Margot said she had developed a conviction that her husband, Gregory, was not interested in her. When she came home from work, Gregory never asked anything about her daily events, and he never told her about his own unless something extraordinary happened. By the time she found out, with the help of a therapist, that Gregory was not uninterested, but *merely silent* and passive by nature, it was too late to save their marriage.

Paul and Michael had been good friends in body and soul when they were at university, and as physics graduates they started working at the same research institute. However, Paul soon won a large international grant, and Michael said he felt left out. The "good news" came out on a Tuesday, Michael stopped saying hello to Paul the next day, and after a few weeks, Michael left the institute.

With these stories in mind, we next discuss some warning signals that might have been identified early on. It might be useful to remember them when we are in troubled relationships.

3.4.3   Warning Signals

If interests diverge, if less time is spent together, you may begin to suspect that there is some problem with your relationship. For example, those who don't watch the same movies, read the same books, have different musical tastes, or have different attitudes toward sports might weaken their relationship.

An obvious pattern is that after a rising phase, the relationship reaches *cruising altitude*. During an eight-hour international flight, a plane spends about seven hours at this height. At the beginning, both partners put a lot more energy into the relationship, looking for opportunities to please the other. As time passes, the enthusiasm diminishes. They reach cruising altitude. The relationship should be maintained.

Let's stick to the parallel of maintenance: In the case of the car, we pay special attention to making sure that the engine, brakes, and other essential equipment work well. Which "parts" of an emotional connection should be given the most attention for maintenance? According to Jennice Vilhauer, respect, love, support for each other and each other's goals, and mutual responsibility are the cornerstones of healthy relationships, none of which can be damaged without risk [25]. In the case of love, the emphasis is not on emotion but deeds. The love you feel for the other is expressed in the way you treat him or her—in kindness, commitment, tolerance, support, benevolence, and even forgiveness or admiration.

When attention to each other decreases, or good conversations are replaced with more and more shouting, threatening, accusing, or quarreling because each or both have become indifferent, these too are signs of a deteriorating relationship. Everybody needs deep and honest conversations based on curiosity, interest, and support. A desire to know each other fills the relationship with passion. It indicates something is amiss if the couple members are less and less interested in each other, if they feel that they do not have anything to say to each other or cannot share what they think or feel with each other. But with mutual attention to one another, they can feel that they are essential to each other.

Another indicator of the deterioration of a relationship is the emergence and prevalence of mistrust. Mistrust can, in many cases, also come from one party's early, negative childhood experiences. But suppose the relationship used to be full of trust but is now filled with silences, gossip, lies, violations of privacy, suspicion, and interrogation. In that case, it reflects a loss of trust. Mistrust can also be based on insecurity and can propagate through a relationship like a wildfire. Wildfire is a natural disaster to be discussed in the next chapter, but it seems a good analogy for this context. First, you may feel some vague uncertainty about the trustworthiness of your partner. Doubt will soon be converted into suspicion over time. Suspicion is nothing but belief without proof.

John Gottman, one of the world's best-known psychologists, identifies four telling signs of deteriorating relationships [26]. These four indicators, also known as the four horsemen, are criticism, defensiveness, contempt, and stonewalling. Please note, conflict itself is not unhealthy! Instead, it might be productive in clarifying the expectations. The "contract" between partners should be renegotiated occasionally. A statement like "you are so lazy" might be the reflection of a desire like "please help more around the house." Defensiveness leads not to the acceptance of any responsibility but to the rapid blaming of the other. The third indicator is contempt. Rolling your eyes or using "humor" to degrade your partner is a typical way to express some feeling of superiority. The last and most dangerous indicator is stonewalling. It means that you are non-responsive and ready to pull yourself out of the relationship.

Gottman studied the stages leading to separation and identified the following features:

- When the couple evaluates that their problems are serious
- If they can't discuss things together and try to solve problems alone
- If they are living a parallel life
- When loneliness becomes a feature of their lives.

In many cases, communication is the best way to repair the relationship …if it needs to be repaired. For some people, this may seem like a superficial treatment, a change of appearance, but we often find that a change in attitude alone can create results. When communication becomes positive and safe, we can approach the more minor or more extensive problems in the relationship with more confidence. After all, communication is a proven way of solving problems in human groups. If it doesn't work, it's like fixing a fence with a broken toolbox or a hammer with a loose handle.

There are many ways to improve communication. It is a strategic issue that we must always take care of: Getting our message across to the receiver and ensuring that he or she does not misunderstand it. We can confirm this by asking questions, so another critical element of good communication is feedback and regularity. If communication is established in a relationship, there is a much lower chance of conflicts escalating, as they can be resolved or talked through in good time.

Gottman also mentions the concept of set point. His analogy is that we have a basis weight that our body is trying to maintain, and because of homeostasis, our body maintains that weight whether we diet or not. (Homeostasis will be explained in Section 4.1.) Only by changing the body's metabolism can we achieve significant weight loss through, for example, dieting and/or taking up regular exercise. Gottman says about marriage that if you set a high baseline emotion, it takes more negativity to damage the attachment than if it is set at a low baseline. In the latter case, it is much harder to improve. Gottman found that most of the couples who attended his workshops were relieved to hear that almost all marriages have conflicts and quite serious ones. The question is whether attempts to repair marriages can be successful. In other words, the key to reviving a relationship is not how the parties handle conflict but what they do when they are not fighting.

We don't always pay much attention to our relationships, so we don't recognize the warning signs of trouble until it's too late. Some relationships can also be intentionally ruined. There are several liquidation strategies. Silent sabotage is relatively passive: "I'd rather not do anything; what will be will be." Provocation, such as overturning an earlier agreement, is an active strategy to destroy the relationship systematically. It is an explicit or implicit, conscious or unconscious sign of considering breaking up.

Warning signals are one of the responses to perturbations that can occur in a relationship, friendship, or employment. If they become regular, we need to pay attention to what they are about—what is missing in the relationship. The relationship has not only a past but also a future. If we don't know about each other's plans—tomorrow, next week, next year—it is a sign that their future may involve separating from ours.

To speak more positively, we should think about saving a struggling relationship. Remember: It is essential to know when it's best to let it go! If you ever feel physically or psychologically threatened, these are absolute red flags that you should not ignore. Relationships can be saved only *if both partners* are willing to work on rescuing the relationship.

So far, we have discussed individual relationships. But many social interactions occur in group settings, and group organization is a fundamental feature of human society. How do social dynamics in larger settings deteriorate?

3.4.4 Destruction of Groups: Dissolution and Polarization

One extreme mechanism of destroying small or large groups is polarization. In a highly cited paper, political scientists have studied the phenomenon of *affective polarization*. More precisely, they analyze its origin and consequences [27]:

> While previously polarization was primarily seen only in issue-based terms, a new type of division has emerged in the mass public in recent years: Ordinary Americans increasingly dislike and distrust those from the other party. Democrats and Republicans both say that the other party's members are hypocritical, selfish, and closed-minded, and they are unwilling to socialize across party lines. This phenomenon of animosity between the parties is known as affective polarization.

The tendency for partisans to dislike and distrust those from the other party is one of the most striking developments of twenty-first-century politics and can also be seen in a good number of European countries. How might the pandemic influence this tendency?

There is a belief that the COVID-19 pandemic may be driving a dramatic increase in social and political polarization. As we discussed previously, positive feedback mechanisms amplify differences, and existing social differences escalate. People with higher educational degrees who predominantly work in office-type jobs rapidly transitioned to their home offices, while others experienced a reduction in working hours or job loss. A third cluster, namely those who are labeled *essential workers*, continued manual labor with many social interactions and were more easily exposed to the virus. The crisis amplifies the inequalities between poorer and richer regions [28].

One possible outcome of the lingering pandemic is that one cluster of people will be ready to follow the recommendations of public health authorities, and the other will resist. There were ongoing debates at the end of 2021 about whether or not mandatory vaccination was a possible institutional policy. It appears that "no major constitutional or international court has found that a mandatory vaccination policy violates any general right to liberty" [29].

In the case of increasing polarization, the role of *bridges* is very important. In network theory, a bridge connects otherwise disjoint clusters. The destruction of bridges leads to the emergence of isolated communities that are not able to communicate with each other. Repairing the world might happen with the reconstruction of bridges, which we discussed at the beginning of this book in Sect. 1.1.1. We also learned in Sect. 1.3.1 how weak connections play an important role in tying together distant clusters.

3.5 Lessons Learned and Looking Forward

The hard laws of physics and the softer laws of social science suggest the malleability of things and connections. The first and second laws of thermodynamics relate to constancy and change. Processes are irreversible, and basically everything is the subject of decay due to wear and tear. Both physical objects and people can be subject to wear and tear. Specifically, burn out is a very extreme form of this phenomenon. We just may hope that the newer generations will learn some mechanism so as not to be the victims of their ambitions in their workplaces.

As concerns societal disasters, we should be prepared for the worst case scenario. By and large, many of us intuitively feel that the short-term thinking of the majority of politicians makes long-term planning impossible, and the myopic perspective may easily lead to existential risk [30]. We discussed possible mechanisms of societal collapse and, in a broader context, of the generation of extreme events. Uncompensated positive feedback might lead to explosion, both in the specific and the more general sense of the word.

Stable relationships are fundamentally important, and we discussed the warning signals evident in the breakdown of relationships. Relationships can be saved, but only if each partner is ready to work on it. The main principle of repairing a relationship is the same as repairing other things of a different nature: "An ounce of prevention is worth a pound of cure," as Benjamin Franklin (1706–1790) once famously said. If we monitor our relationships regularly, even when everything is fine (or seems to be fine), we can spot small things that might become problems over time. Proactively dealing with these can prevent them from growing bigger and escalating or even tearing the relationship apart.

We can prevent the breakdown of relationships by identifying the transitions that are likely to occur. Today, social scientists are studying and publishing on these topics to help individuals better preserve the health of relationships of all kinds. Zsuzsa knows a couple in their sixties who go on holiday with their grandchildren, among other things, so that their daughter and her husband can spend time alone every year, preserving the intimacy of their relationship. Some managers regularly ask their employees how they find their tasks and whether they can improve in their work. Retention and renewal are the cornerstones of any relationship. They are coupled with our fundamental, sometimes contradictory, needs to be both independent and part of a community or team. If we are aware of this and know, for example, that love tends to change after two or three years, or have read about the empty-nest syndrome experienced by middle-aged parents, we can apply the knowledge available today to prevent relationships from deteriorating.

Murphy's law famously states that "If something can go wrong, it will." Industrial catastrophes almost always originate from human error, through lack of attention, communication, or competence. We humans do not like to admit our mistakes and like to find excuses from software bugs to acts of God, often blaming things on Murphy's Law. Can we avoid Murphy's law? Accepting personal responsibility for something might work better than blaming others. Nonetheless, industrial designers have realized the necessity of constructing fail-safes or "idiot-proof" systems to minimize the risks of human error. We need an "idiot-proof" political system to survive.

To be a little bit more specific, we need a transition from the Throwaway Society to the Repair Society. The next chapter studies the question of how we can restore disturbed states, at different levels, from objects to buildings to smaller and larger communities [31–33].

References

1. Snow CP (2001) The two cultures. Cambridge University Press, p 3 (Original work published 1959)
2. Fabian D, Flatt T (2011) The evolution of aging. Nature Educ Knowl 3(10):9. https://www.nature.com/scitable/knowledge/library/the-evolution-of-aging-23651151/
3. Mitchell SW (1897) Wear and tear, or. Lippincott, Hints for the overworked. J.B
4. Davison R (2019) Toughened glass exploded 'like a grenade.' Otego Daily Times. https://www.odt.co.nz/regions/south-otago/toughened-glass-exploded-grenade
5. Rupert ML (2013) Spontaneous glass breakage: why it happens and what to do about it. The Construction Specifier. https://www.constructionspecifier.com/spontaneous-glass-breakage-why-it-happens-and-what-to-do-about-it/
6. Petersen AH (2019) How millennials became the burnout generation. BuzzFeed News. https://www.buzzfeednews.com/article/annehelenpetersen/millennials-burnout-generation-debt-work
7. Han BC (2015) The burnout society. Stanford University Press
8. Maslach C, Jackson SE (1981) The measurement of experienced burnout. J Organizational Behav 2(2):99–113. https://doi.org/10.1002/job.4030020205
9. Aumayr-Pintar C, Cerf C, Parent-Thirion A (2018) Burnout in the workplace: a review of data and policy responses in the EU. European Foundation for the Improvement of Living and Working Conditions. https://rhepair.fr/wp-content/uploads/2018/11/2018.09-Burnout-in-the-workplace-A-review-of-data-and-policy-responses-in-the-EU-Eurofound.pdf
10. Spring Health (2020) Study finds 76% of U.S. employees are currently experiencing worker burnout. PR Newswire. https://www.prnewswire.com/news-releases/study-finds-76-of-us-employees-are-currently-experiencing-worker-burnout-301191279.html
11. Maxmen A, Mallapaty S (2021) The COVID lab-leak hypothesis: what scientists do and don't know. Nature. https://www.nature.com/articles/d41586-021-01529-3
12. Bak P (2007) How nature works: the science of self-organized criticality. Copernicus
13. Sornette D (2003) Why stock markets crash: critical events in complex financial systems. Princeton University Press
14. Cowen T (2020) Is economic growth a moral imperative? Discourse Magazine. https://www.discoursemagazine.com/culture-and-society/2020/03/05/is-economic-growth-a-moral-imperative/
15. Raworth K (2017) Doughnut economics: seven ways to think like a 21st-century economist. Chelsea Green Publishing

16. Maldini I (2021) The Amsterdam doughnut: moving towards "strong sustainable consumption" policy? Presented at 4th PLATE 2021 virtual conference. https://ulir.ul.ie/handle/10344/10228
17. Meadows DH, Randers J, Meadows DL (1972) Limits to growth. Penguin Publishing Group
18. Sornette D (2002) Predictability of catastrophic events: material rupture, earthquakes, turbulence, financial crashes, and human birth. Proc Natl Acad Sci U S A 99:2522–2529. https://doi.org/10.1073/pnas.022581999
19. Sutala T (2004) Mechanisms of epilepsy progression: current theories and perspectives from neuroplasticity in adulthood and development. Epilepsy Res 60(2–3):161–171. https://doi.org/10.1016/j.eplepsyres.2004.07.001
20. Ben-Ari Y, Crepel V, Represa A (2008) Seizures beget seizures in temporal lobe epilepsies: the boomerang effects of newly formed aberrant Kainatergic synapses. Epilepsy Curr 8(3):68–72. https://doi.org/10.1111/j.1535-7511.2008.00241.x
21. Frolov NS, Grubov VV, Maksimenko VA, Lüttjohann A, Makarov VV, Pavlov AN, Sitnikova E, Pisarchik AN, Kurths J, Hramov AE (2019) Statistical properties and predictability of extreme epileptic events. Sci Rep 9:7243. https://doi.org/10.1038/s41598-019-43619-3
22. Golestani A, Gras R. Can we predict the unpredictable? Sci Rep 4:6834. https://doi.org/10.1038/srep06834
23. Dunbar R (2021) Friends: understanding the power of our most important relationships. Little, Brown Book Group
24. Argyle M, Henderson M (1984) The rules of friendship. J Soc Personal Relationships 1(2):211–237. https://doi.org/10.1177/0265407584012005
25. Vilhauer J (2017) 4 signs that it's time to get out of your relationship. Psychology Today. https://www.psychologytoday.com/intl/blog/living-forward/201712/4-signs-its-time-get-out-your-relationship
26. Gottman J, Silver N (2015) The seven principles for making marriage work. Harmony Books
27. Iyengar S, Lelkes Y, Levendusky M, Malhotra N, Westwood SJ (2019) The origins and consequences of affective polarization in the United States. Ann Rev Political Sci 22:129–146. https://doi.org/10.1146/annurev-polisci-051117-073034
28. Jungkurz S (2021) Political polarization during the COVID-19 pandemic. Front Political Sci 3:622512. https://doi.org/10.3389/fpos.2021.622512
29. King J, Ferraz OLM, Jones A (2021) Mandatory COVID-19 vaccination and human rights. The Lancet 399(10321):220–222. https://doi.org/10.1016/S0140-6736(21)02873-7
30. Leigh A (2021) What's the worst that could happen? MIT Press
31. Érdi P (2007) Complexity explained. Springer, Heidelberg
32. Diamond J (2005) Collapse: how societies choose to fail or survive. Penguin Books
33. Kim J (2017) The most important factor in successful relationships. Psychology Today. https://www.psychologytoday.com/us/blog/the-angry-therapist/201706/the-most-important-factor-in-successful-relationships

Chapter 4
The Pathways Back to "Normal"

Abstract In this chapter we reviewed basic mechanisms for responding to disturbances, which can drive systems back to their original states. It looks Nature provides
some compensatory mechanisms to promote adaptation to the continuously changing environment. The observation, which became known as Le Chatelier's principle,
speaks about the stability of a thermodynamic equilibrium state, but its spirit was
soon extended to to biological and social systems too. Homeostasis should be considered as a general repair mechanism that ensures the functional stability of living
systems. Living systems are in continuous interaction with their environment through
material, energetic, and informational flows. The stable operation of biological and
social systems is maintained by the balance of positive and negative feedback loops.
When the balance is violated, huge systemic failures, such as social inequality and
climate change, emerge. The concept of resilience originated from the field of ecology, but its applicability and significance has escalated in light of natural and social
disasters. A system is resilient if it continues to carry out its function in the face of
adversity. We discussed the importance of being resilient at many levels of hierachical
organization, from individual to buildings to small and large communities.

4.1 Stability, Homeostasis, and Resilience

4.1.1 Stability

In this chapter, we focus on mechanisms by which things get better—more precisely,
how they get to be as good as they were before they were impaired.

There is a general mechanism for restoring certain states after a small perturbation,
as illustrated in Fig. 4.1. These states are called *stable* equilibria, and while their
definition comes from physics and chemistry, the concept can also be applied to
social systems too, as the list of examples below indicates [1]:

- A ball contained in a cup that is struck very hard will return to its original position
- A candle flame diverted by a slight puff of air will regain its original form
- A trout brook that is "fished out" will, if carefully protected, regain its former
  population of fish

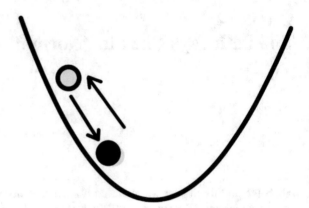

Fig. 4.1 If a small perturbation is applied, the system returns to the equilibrium state

- Children that lose weight because of a mild illness will catch up with the weight they might have been in the absence of sickness
- Economies exhibit some stable equilibria, showing adaptations to small changes— if supply and demand are in equilibrium, they determine the equilibrium price, and small disturbances generate economic pressures that move the market toward equilibrium.

The simplest mechanism of repair is the one at work when a system can be restored, assuming any disturbance is small. Such systems are called *stable*. Too much stability, however, can work against adaptation to environmental changes. The trade-off between stability and adaptability is often called the *stability-plasticity dilemma*. We will return to this concept after first discussing mechanisms of restoration observed at the minute level of chemical processes.

Le Chatelier's Principle and Its Application in Nature and Society

Le Chatelier's principle is an observation about the chemical equilibria of reactions. It states that changes in physico-chemical conditions—such as temperature, pressure, volume, or concentration of a system—will result in opposing changes within the system to achieve a new equilibrium state. Once such a principle had been articulated, many scientists felt that its spirit might be valid in explaining many other systems in nature and society. More generally, when external stress is applied to a system in equilibrium, the equilibrium will change in such a way as to reduce the effect of the stress. In other words, a change in a system will evoke a counter-change, which will bring the equilibrium to a new point. Specifically, Paul Samuelson (1915–2009) extended the principle to economics. Demand, supply, and price substitute for physical state variables, and the theory explains how the market responds to changes in supply and demand [2].

COVID-19, Other Disasters, and Le Chatelier's Principle

As we experience the pandemic and other disasters, scientists have adopted analogies and models to understand these phenomena better. For example, the clinical biochemist Eleftherios P. Diamandis applied the spirit of Le Chatelier's principle to argue that the greediness of the human population provokes responses like wildfires, tsunamis, floods, extreme heat, and new diseases and pandemics: "The more we abuse the planet, the more it will resist and pay us back with interest" [3]. SARS, MERS, and Ebola were strong warning signals. A hundred years ago, the Spanish flu killed 50 million people, but …humans have short-term memories.

The Lesson of Le Chatelier's Principle
"If we continue to perturb the equilibrium on the planet we should be ready to face the inevitable counter-changes and pay the associated price." [3]

Homeostasis

Living systems defend themselves against permanent changes in the environment, and we may consider the means by which they do so a basic repair mechanism. The French physiologist Claude Bernard (1813–1878), the founder of modern experimental physiology, was perhaps the first to fully appreciate that living systems possess characteristics of internal stability that buffer and protect the organism against a constantly changing external environment. Homeostasis should be understood as a general repair mechanism in response to environmental changes [4]. An organism is not a static, but a dynamic, self-regulating system. Constancy or stability is maintained through the interaction of negative and positive feedback loops, which we discussed in the previous chapter.

In his book, *Cybernetics*, Norbert Wiener (1894–1964) developed the first formal mathematical analysis of feedback control in biological systems [5]. Models of physiological systems extensively apply control theory, but even more importantly, Wiener generalized the concept of homeostasis from the individual level to the societal level. *Cybernetics*, the scientific discipline for which Wiener named his book, describes these relationships and applications. The subtitle of the book was "Control and Communication in the Animal and the Machine." While the physiologists already knew that the involuntary (autonomous) nervous system controls the internal environment (i.e., Bernard's famous "internal milieu"), Wiener extended the concept and suggested that the voluntary nervous system may help maintain the external environment through feedback mechanisms. The theory of *goal-oriented behavior* provided a new framework for understanding the behavior of animals, humans, computers, and a rapidly transforming society.

Learning and the Stability-Plasticity Dilemma

The "constancy versus change" dichotomy appears in a somewhat different form in learning systems. We somehow balance between absorbing a vast amount of new information and forgetting some items. Learning happens in biological systems, but artificial intelligence and machine learning have become hot topics in the age of the data deluge. We can rapidly absorb a vast amount of new information throughout our lifetimes. And humans are able to integrate all of this information into a unified conscious experience. As Stephen Grossberg, one of the pioneers of modern brain theory wrote, "One has only to see an exciting movie just once to marvel at this capacity, since we can then tell our friends many details about it later on, even though the individual scenes flashed by very quickly" [6, 7].

More generally, we can quickly learn about new environments, even if no one explicitly tells us how the rules of each environment differ. To a remarkable degree, humans can rapidly learn new facts without being forced and, just as rapidly, forget what we already know. As a result, we can confidently go out into the world without fearing that, in learning to recognize a new friend's face, we will suddenly forget the faces of our existing family and friends. This property is often called "catastrophic forgetting." Grossberg's "Adaptive Resonance Theory" has been suggested as a model of the neural mechanism behind the brain's ability to autonomously categorize, recognize, and predict objects and events in a changing world. For our purposes, this means we have a neural mechanism that explains our ability to maintain and update our complicated memory systems.

4.1.2   Feedback Loops Everywhere

The thermostat is a device that implements *balancing feedback*. Its job is to balance the system (your home) around the desired state (a comfortable temperature). A thermostat is able to do two things: It adds heat when there's too little and shuts the heating down when there's too much. It employs a negative feedback loop since it is acting in opposition to forces imposed on the system (if you open a window and the temperature drops, the thermostat causes the heating to turn on—that's the opposite of what would otherwise happen were the system to reinforce the change). We introduced the concept of uncompensated positive feedback in Sect. 3.3.1. As the current topic is restoring mechanisms, let's see more examples.

From Escalation to Agreement

Each year I (P) play a game with my students. I ask one of them to answer my initially quiet "Yes," with a louder "No!" My answer is a louder "Yes!" followed by her stronger "No!" Finally after a shouted "Yes!" and "No!", a frightened colleague looks from the corridor into the lecture room and seems to be happy that there is no bloodshed. Any of us knows this phenomenon. I hit you, and you hit me harder, so there is a process of mutual activation. Ending the fight requires some kind of damping mechanism. Among people, so too among states: History teaches us about inter-state wars and the existence of *non-aggression pacts*, a mechanism for restoring peace. States with a history of rivalry might sign non-aggression pacts to prevent future conflict with one another. The deals are often meant to increase information exchange, which reduces the uncertainty that sometimes leads to war.

Success Makes You Successful

In his book about the formulae of success, Albert-László Barabási explained the amplification of minor initial differences: Papers with more early citations had a better chance of being cited again than papers with fewer early citations [8]. (His initial fame derived from the significant number of citations generated by his discovery of the "preferential attachment" mechanism in the evolution of the World Wide Web.) When you get a grant that enables you to get better equipment or hire more coworkers, this improves your odds of winning another grant.

Inequality of Wealth: How to Stop the Runaway Process and Return to Homeostasis

Thomas Piketty's bestseller, *Capital in the Twenty-First Century*, argues that economic inequality in American society (and possibly others) is even worse than people believe, and it will continue to worsen under the conditions of present-day capitalism [9].

Data suggest that while in the 1970s, the top 0.1% of American families owed "only" 7% of national wealth, this figure has risen to about 25%. Piketty shows that the basic feature of present-day capitalism is that the rate of return on capital (r) is greater than the rate of economic growth (g). Under the condition $r > g$, wealth will tend to concentrate in the upper minority. It is clear that in America's current economic system, the "rich get richer" mechanism is at work. While the average CEO's pay increased by 56% over the course of a decade, the median worker's pay increased only 18%, leading to an executive-to-worker pay ratio of 287 to 1 in 2018 [10].

Moreover, economic power can be converted into political influence, accelerating the runaway process. The most obvious stabilizing negative feedback could be implemented by redistributing wealth through instruments such as progressive taxation. We might hope that such interventions could restore social homeostasis, as complex systems theory suggests. While the billionaire tax suggested by the Democrats during the fall of 2021 would heavily target the wealthiest Americans, the plan is not likely to come to fruition. As it stands now, in a split Senate, one moderate Democratic Senator can veto the plan and render it dead.

Of course, the emergence of inequality is a complicated process. But we have a very good *skeleton model* by taking into account an *accumulation process* generated by a positive feedback mechanism, potentially compensated by some *redistribution process* to ensure stability.

Somewhat similarly, the positive feedback mechanism responsible for climate change has been identified, but it is not clear how negative feedback will be implemented to avoid a climate crisis.

Feedback Loops and Climate change

Climate change, climate crisis—we hear these terms a dozen times each day. Climate theory defines three groups of factors that influence climate change: (1) forcing, (2) feedback loops, and (3) tipping points, as a NASA report suggests [11].

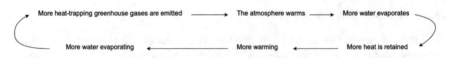

Fig. 4.2   An example of a positive feedback effect in climate change is the water vapor cycle

Examples of forcing are identified as the initial drivers of climate. The most important drivers are:

- Solar irradiation
- Greenhouse gas emissions
- Aerosols, dust, and smoke.

As the NASA report also suggests, among the three drivers, the last two are anthropogenic (i.e., they are human products). The concentration of greenhouse gases in the atmosphere has increased exponentially (so, maybe not yet super-exponentially), causing increased warming of the atmosphere and global climate change (Fig. 4.2).

Without negative feedback to stop this vicious cycle, the system can reach a tipping point. Seemingly small changes might result in dramatic, irreversible changes affecting ocean circulation, ice loss, or rapid release of methane.

There are some moderately optimistic statements that we are not yet there. Even Nobel Peace Laureate and former U.S. Vice President Al Gore's Climate Reality Project writes on its blog "there's still plenty we can do to limit the worst impacts of the climate crisis and support a safe, sustainable future" [12].

4.1.3   Resilience

What Are We Talking About?

There is another magic word: *resilience*. C.S. Holling (1930–2019), one of the founders of ecological economics, articulated the concept of resilience in the field of ecology in 1973 [13]. Resilience refers to a measure of the persistence of systems and of their ability to absorb change and disturbance and still maintain the same relationship between populations or state variables: "Stability represents the ability of a system to return to an equilibrium state after a temporary disturbance; the more rapidly it returns to equilibrium and the less it fluctuates, the more stable it would be." Holling distinguished between two types of resilience: one in which a system can return to a steady-state following a perturbation, and another in which the disturbance causes the system to transform into a new state. (Recently, this has been referenced in popular conversations about the "new normal," which we will discuss in Chap. 5.)

Resilience is a multidisciplinary topic spanning the natural sciences, social sciences, and engineering and has a broad set of applications from individual to household, communal, regional, and global levels. Its goal is to answer questions of how to restore livable conditions after personal tragedies and natural and social disasters.

Specifically, where climate change is concerned, climate hazards can cause health impacts ranging from mild injury to loss of life and damage to infrastructure, property, services, and other environmental resources. Even before the pandemic, about 100 million people were affected annually by weather- and climate-related natural disasters across the globe [14]. Many regions in Africa and South/Southeast Asia have the highest risk, since they suffer from a relative lack of financial resources to prevent and cope with climate impacts. We should understand that the seriousness of disasters we call "natural" is partly the result of insufficient planning and preparation.

Some (Not Only) Recent Disasters

Heat waves can emerge as a result of several factors, including urban design, materials with high thermal mass, low albedo, and low permeability. (Albedo is a measure that characterizes a surface's ability to reflect solar energy. It is 0 for black and 1 for white.) In recent years, St. Louis, Missouri, has had huge heat waves, as has been widely publicised [15]. But actually, it is not so new! In the summer of 1936, "St. Louis endured an unbroken 37-day stretch of 100-degree–plus temperatures."

While *drought* occurs all over the world, the eastern Horn of Africa is hit more frequently than other areas. From a naive perspective, it is surprising that even Cape Town, which has beaches on both the Atlantic and Indian Oceans, experienced a prolonged water crisis. Between 2015 and 2018, city management had to implement significant water restrictions to deal with the drought. However, the system displayed some resilience, and with sufficient rain in 2020, the dam level was functionally restored. The city of Chennai in India experienced a similar drought in 2019 [16].

Drought is often followed by heat-induced *wildfires* in Australia. Approximately 18.6 million hectares of land burned through, destroying 6000 homes over the summers of 2019 and 2020. This wildfire was rather persistent, but it was successfully controlled after eight months of intervention. Predictive models suggest that the probability of wildfire has increased quite dramatically.

There are a good number of similar examples, including flooding and increased sea level, which will have a major impact on buildings and people at the individual, community, and regional levels. Having briefly surveyed some climate-related disasters, we now turn to discuss resilience at these different levels.

Resilience at the Individual Level

In general, we are aware when a particular impaired relationship or situation needs to be repaired. Some painful events may dominate our whole self. After a series of particularly negative tragedies and traumas, we may feel that our *whole* life is broken and needs fixing. Trying to repair a specific problem may increase our resilience. How?

Anthony, one of Zsuzsa's colleagues, adopted some useful strategies after his divorce. He found break-up diaries and exercises online: He used them to train himself until he got through the most difficult months. In addition, he watched a lot of movies about divorce and in almost all of them he discovered something that helped him interpret his own situation. As Anthony analyzed his divorce—took it apart and put it back together—he did the same for himself, in a sense. He came out of it stronger and more resilient than when he went in.

Edith's beloved brother died unexpectedly in an accident. Someone in her life suggested she join a grieving support group. The helpful influence of a supportive community is one of the most powerful healing forces. Edith was afraid at first, thinking that her grief would be magnified if she had to share in the grief of others, but just the opposite was true. Edith's pain soon eased because she experienced that in the group she could remain silent when she was embarrassed but could also talk openly about feelings she would initially have been ashamed to express in front of others. The group was understanding of Edith's sense of injustice for the tragedy that had happened to her, and the insight she gained into the feelings and thoughts of other people in similar situations helped her.

Both Anthony and Edith became more resilient by the "end" of the healing process. The take away is that anyone can learn the behaviors and thoughts that lead to resilience. It is not an extraordinary but rather a very common personality trait. Negative changes affect everyone differently, but the flood of insecurities and painful emotions and thoughts is always unique. Yet most people are able to adapt to stressful situations over time—to develop resilience [17].

Resilience is about adapting, bouncing back, and developing personally in a profound way. There are always more aspects of our lives that we can control or modify and thus develop, improve, and grow with them.

Some strategies and actions work for the vast majority of people. We need willingness, time, and practice, but we also need trustworthy companions who validate our feelings. Our relationships and communities not only protect us from isolation, but they are also a source of support and joy.

We also need self-care. After all, stress is physical, not just emotional, so a positive lifestyle (from sport to proper sleep to healthy eating) can reduce the burden of anxiety. Mindfulness (i.e., a type of meditation in which you focus on being intensely aware of what you're sensing and feeling in the moment) has an increasingly good reputation among millennials, but it needs more scientific research to make sure it is not over-hyped [18]. We can also take care of ourselves by taking care of others. Volunteering, for example, can boost our self-worth by helping others in a tangible way: alleviating the hardship of other people in need.

Not all negative events can be changed. But we can change the way we interpret traumas, the way we react to them, and the way we look for resources to solve them, recalling the experiences of previous difficult situations. Do we accept what is unchangeable from the situation or do we see it as a catastrophe? The Serenity Prayer written by Reinhold Niebuhr (1892–1971) is still widely applicable [19]: "God, grant me the serenity to accept the things I cannot change, courage to change the things I can, and wisdom to know the difference."

Resilience is the ability to adapt to difficult situations, trauma, and severe adversity: When we experience pain or anger, we remain functional, both physically and psychologically. We can cope better with stress, move beyond the problem at hand, find joy in life, see the future, and have hope. Vision and hope also help us become proactive, make plans, and prepare for anticipated situations [20].

A few more words about resilience at the personal level: Recent commentators have noted the concept of *resilience inequality*. The COVID-19 crisis made it very visible that, for "rich" people, the economic effect was at most a temporary shock. By contrast, those without savings cannot cope with even a relatively small loss without fearing severe economic consequences. The resilience inequality might be transferred to income and later wealth inequality.

Resilience at the Building Scale

In preparation for increasingly severe weather, architects, engineers, and urban planners should ask and answer how we can *design* and construct or renovate buildings to cope with flooding, severe storms, increased sea levels, wildfire, and other extreme weather events.

The United Nations Environment Programme offers a practical guide to climate-resilient buildings and communities [21]. But what does it mean for buildings to be resilient? A building should have the ability to provide safe, steady, and comfortable use even when outside conditions change. Such changes may occur due to drought, flooding, extreme precipitation, or heat. (The spirit of Le Chatelier's principle works here too: If a change produces discomfort, people tend to react in such a way as to restore their comfort.)

Many big earthquakes have destroyed buildings and people over the course of Japan's history. While the Tokohu earthquake of 2011 is known as one of the most devastating in recent times, it was one of many seismic events that strike Japan each year. As a consequence, there are rigorous rules in place for designing resilient buildings—even if they are minor or temporary structures. Japanese engineers set two thresholds. First, minor earthquakes, which occur three or four times in a building's lifespan, should not result in any damage requiring repair (in other words, buildings should be sufficiently well-designed to avoid any functional loss). Second, in the event of rare, extreme earthquakes, buildings should be designed with the goal of avoiding human casualties. To achieve this goal, buildings should be able to absorb as much seismic energy as possible. Blocks of rubber may serve to provide seismic isolation. A tall building may move 1.5 m, but placing dampers, say every second floor, strongly reduces the probability of damage. In the case of Tokyo Skytree, now the second-tallest structure in the world, designers believe that half of the energy from an earthquake can be absorbed by such dampers. This technology enables the present trade-off between the height and resilience of tall buildings.

Resilient buildings should, among other things:

- Maintain livable conditions in the event of extended loss of power or heating fuel
- Be built in terms of water conservation and do things like harvest rainwater and use it as the primary or back-up water supply
- Provide redundant water supplies or water storage for use during emergencies, including hand-pumping options where possible

- Include an option for human waste disposal in the event of non-operating municipal wastewater system, which could include composting toilets and waterless urinals
- Provide redundant electric systems with at least minimal back-up power capacity, such as a fuel-fired electric generator (with adequate fuel storage) or a solar-electric system with islanding capability
- Maintain non-perishable food supply that could provide residents with adequate staples for a 3-to 6-month period (e.g., canned goods or dehydrated foods like dried fruits and vegetables).

Resilience at the Community Scale

How should we design resilient communities? Both physical and social factors must be taken into account. Physical infrastructures are composed of transportation systems (roads, bridges, tunnels, ports, rail, airports) and utility plants and distribution systems (electric power, water and waste-water, fuel, communication systems). Communities should build social structures that strengthen the fabric of the community. Community gathering places, dog parks, central mailbox locations, and community bulletin-boards are examples. Resilience hubs may play an important role in case of emergencies, or the interruption of certain services, by providing water, electricity for charging cell phones, and other goods and services. Public education systems should include programs that promote an understanding of energy, water, and other natural resource systems, as well as the functioning of buildings and community infrastructure.

In 2020, the United Nations launched the Making Cities Resilient 2030 (MCR2030) initiative. In 2021, four European cities (Barcelona, Greater Manchester, Helsingborg, and Milan), were named the first "Resilience Hubs" for their policy and vision in addressing growing climate and disaster risks.

The Japanese "Koban" may also provide a useful model. Japan is known as one of the safest countries in the world. One significant factor that keeps Japan peaceful is the institution called "Koban," which is a "police box." (This concept evolved from that of the telephone box.) In addition to keeping the neighborhood quiet, Koban has other roles. Police officers at Koban take care of children who get lost, they give directions to people who cannot find their way, they keep lost property for owners to collect, and they are ready to listen to the worries of local residents.

Resilience at the Regional Scale

Resilience at the regional scale is one of the most effective ways to help communities adapt to climate and economic changes. Specifically, "resilient cities, those that are working to transition towards a low-carbon economy while also preparing to avert the worst of climate change, are gaining interest and attention from policy makers, city councils and others worldwide" [22]. Like it or not, ranking is with us [23]. In 2014, Grosvenor—a property investment company—published a research report rating cities according to their vulnerability and capacity to adapt to environmental and other threats. Here is a ranking of the Top 10 Most Resilient Cities in the world [24]:

1. Toronto
2. Vancouver
3. Calgary
4. Chicago
5. Pittsburgh
6. Stockholm
7. Boston
8. Zurich
9. Washington, D.C.
10. Atlanta.

Please note that ranking is a navigation between objective and subjective, and Canadian cities may be slightly overrated. In 2021, the United Nations named Dubai the world's most resilient city [25].

Resilient Society

In the U.S. context, concerns about resilience have been evident in public discourse and policymaking about supply chain resiliency. In February 2021, President Biden signed an Executive Order on America's Supply Chains, directing federal stakeholders to identify vulnerabilities and develop a strategy to promote resilience in the nation's critical supply chains [26]. The President cited an old proverb: "For want of a nail, the shoe was lost. For want of a shoe, the horse was lost. And on, and on, until the kingdom was lost" [27]. Small failures at even one point in the supply chain can impact jobs, families, communities, and America's security. In the report produced pursuant to the Executive Order, four key sets of products were identified: semiconductors, large-capacity batteries for electric cars, rare earth metals critically important to the defense industrial base, and pharmaceuticals. The report is nationalist in tone and suggests that supply chains can be reliable and that the United States should restore its productive capacity.

Zooming out to look at the global context, the pandemic has left us with the question of how resilient human society really is. Markus Brunnermeier, in his book *The Resilient Society*, suggests that we need a new social contract [28]. There have been both positive and negative developments, and we can't yet tell what will happen, but we should assume that human society needs to be prepared for a future pandemic, cyberattacks, and climate catastrophes. One piece of good news is that science-based vaccine development was very successful. On the other hand, vaccine skepticism is bad news. In the United States, mandatory vaccination plans proposed and implemented at the federal level can be challenged by states. We can hope that local political interests will not overcome science-based rationality. At the end of 2021, the level of vaccination was already sufficient for the global economy to begin to bounce back, but the disruptions to the global supply chain will remain long-lasting. We already see unimaginably sparse shelves in the shops. (Then came Omicron ...)

One lesson that the world should learn is that we don't need to adopt a very pessimistic perspective to imagine that newer and newer natural and social disasters might happen. We should also learn that to defend ourselves, we need to implement a resilient global system, one that is able to bounce back after shocks.

4.2   Back to Normal: Some Case Studies

4.2.1   Recovery from Burn Out

As we discussed in Sect. 3.2.2, burn out appears to be occurring increasingly frequently due to the consequences of external (and more and more often internal) stress. We don't believe there is any "silver bullet" to recover from burn out. Some recommendations focus on the neurophysiological state of the brain, but is reasonable to assume that there are more passive and active recovery strategies.

The typical passive strategy is binge-watching Netflix series and binge-drinking beer. Binge-watching acts like a drug, causing the brain to release dopamine, and at the beginning you feel somewhat better. But it negatively impacts sleep patterns, produces some addictive symptoms, and affects relationships, goals, and commitments. Things like strenuous exercise, yoga, and walks in the forest are active tools [29]. Recovery from burn out does not happen over a weekend. It really needs time. Some guidelines might help [30]:

- Acknowledge that you're experiencing burnout
- Identify what you need to add (and what you need to subtract) from your life
- Reach out to your friends and family members
- Get outside for 20 min a day, five days a week
- Limit your media intake
- Reduce your job demands–and increase your work resources.

4.2.2   The Art of Restoring Buildings

Preserving and restoring old buildings is essential for people who feel it important to connect the present with the past. We measure our successes against those of the past, and the monuments of the past reflect our history. They contribute to our understanding of the different habits and traditions of people who lived in previous eras. Restoration is both a symbolic and a physical act that connects the past to our present and to the future. Our moral responsibility is to repair and preserve historical monuments to honor our ancestors and transfer collective memories to future generations.

Changing Perspectives toward Repair

There is a saying attributed to Adolph Napoleon Didron (1806–1867), a French archaeologist: "For ancient monuments, it is better to consolidate than repair, better to repair than to restore, and better to restore than to reconstruct" [31]. The terminology is somewhat arbitrary, but there have been two extreme schools of thought: On the one hand, the terms *preservation* and *conservation* have been used to refer to the architectural school of thought suggesting that buildings should be protected and maintained in their current state. The goal is to prevent further damage and deterioration. On the other hand, *restoration*, and even moreso *reconstruction*, are associated with the belief that historic buildings could be improved, or even completed, using modern materials, design, and techniques.

Victorian restoration of medieval churches was widespread in England, but the results are now generally regretted. During the reign of Queen Victoria, the Cambridge Camden Society (CCS) and the Oxford Movement, supported by the English Church, organized the restoration of churches, in order to help revive the medieval attitude to churchgoing. The CCS recommended that whole churches should be restored to the best and purest style, of which traces remain from the Decorated Gothic age (from the late thirteenth century). Each medieval church had at least some small remnant of this decorated style. More than 7000 parish churches in England and Wales were restored during the several decades of the mid-nineteenth century. However, the "restoration" mentality created an idealized state of the past in which these buildings had never existed. This attitude may remind us of the desire to return to a Golden Age that never was.

More generally, conservation professionals have traditionally been opposed to reconstruction because this approach can *falsify history* and create fictional places that never existed [32]. There are ongoing debates among experts as to when reconstruction is an acceptable strategy: "In relation to authenticity, the reconstruction of archaeological remains or historic buildings or districts is justifiable only in exceptional circumstances. Reconstruction is acceptable only on the basis of complete and detailed documentation and to no extent on conjecture" [33].

We are now in a new situation in which the destruction of cultural heritage is occurring more frequently due to both natural and social disasters. Whether or not significant monuments should be reconstructed, even "new" architectures will differ from the original. An example of a natural disaster destroying buildings is the 2015 earthquake in the Kathmandu Valley, Nepal, where close to 10, 000 people died. Hundreds of structures listed as World Heritage Sites were affected. The demolition of the ancient Buddha statues in the Bamiyan Valley in Afghanistan by the Taliban in 2001 and the destruction of one of the oldest and best-preserved gems of ancient civilizations in Palmyra and Aleppo in Syria are huge cultural losses.

From Building Diagnostics to Decision Making

Materials science analysis is often based on collected samples. However, in the case of monuments, it is generally forbidden to take samples. Therefore, non-invasive and no-contact practices have been adopted to acquire the necessary information. In addition to visual inspection and infrared thermography, laser scanning and image-based techniques are used to keep the monument intact and allow historians, archaeologists, and others to study its structure. Data can be collected about the architectural, historical, geometric, and building materials, and a combination of computational algorithms and human experts allow the assessment of a structure's current state.

Once such an assessment has been made, we have the usual questions: Where are we now? Where do we want to be? How do we get there? The answers often depend on the philosophical perspectives of the conservation professionals. A possible categorization is as follows [34]:

- Purist view: "The idea that there can be alternative philosophical approaches to the preservation of buildings is seriously misleading. Correctness cannot be watered down."
- Pragmatist view: "A sound philosophy is one which points in the right general direction—that of truthfulness. Its precise application must depend on the building and its circumstances. If I am in command of all the facts, then the building itself will tell me what to do."
- Cynical view: "Conservation is a completely artificial procedure, interfering with natural processes of decay of absolution. Conservation philosophies are therefore necessarily artificial."

The purist and cynical views are probably too extreme. The selection of a repair technique should combine ethical principles with technical possibilities.

United States: National Register of Historic Places

According to the criteria determined by the U.S. Department of the Interior, buildings and other structures in the United States must be 50 years of age or older to be listed in the National Register of Historic Places. The Register contains almost 100, 000 items [35].

Preservation, rehabilitation, restoration, and reconstruction.

- Preservation focuses on the maintenance and repair of existing historic materials and retention of a property's form as it has evolved over time.
- Rehabilitation acknowledges the need to alter or add to a historic property to meet continuing or changing uses while retaining the property's historic character.
- Restoration depicts a property at a particular period of time in its history, while removing evidence of other periods.
- Reconstruction re-creates vanished or non-surviving portions of a property for interpretive purposes.

To give just one example, New York City announced plans to restore the city's historic Chinatown building after a 2020 fire. The present plans combine preservation and restoration of the building's historic façade and the addition of two floors.

We are sure that the Reader remembers well the video of the fire in the cathedral of Notre-Dame in Paris in April 2019. As in a disaster movie, the flames swallowed the cathedral's attic before attacking the roof and the iconic spire. The cathedral is obviously an irreplaceable monument not only to France but also to the whole world. French president Emmanuel Macron declared his intention to have the church repaired not later than July 2024, when the city hosts the 2024 Summer Olympics Games. By that time, people will be able to enter the building, but the restoration will not necessarily be over. Notre Dame is Notre Dame, one of the most emblematic monuments in the world. As of fall 2021, roughly $950 million has been pledged from private and corporate donors to aid in the restoration. About 1000 oak trees donated from both public and private forests will be used in the reconstruction. Scaffolding has been built around the cathedral to restore the spire, but the pandemic caused a three-month delay in the process.

Turning to another part of the world, Japanese castles often reflect their local histories, and notable historical figures often lived there. Attitudes toward these castles have changed over time: At the end of the nineteenth century, castles were viewed as symbols of the dark past, and many of them were demolished. As Japanese militarism strengthened in the 1920s, the castles came to symbolize Japanese military power. Osaka Castle was reconstructed first, and after World War II, many castles were rebuilt. In general, modern materials like steel-reinforced concrete have been used for the restoration of many castles. It seems to be very characteristic in Japan that the love of tradition is combined with contemporary technology. While the outside preserves historical glory, the inside is a modern building with elevators and air-conditioning.

4.2.3 Resilience After Hurricane Katrina: Where Are We Now?

What Happened?

Hurricane Katrina made landfall in Louisiana and other regions of the Gulf Coast of the United States on August 29, 2005. The day prior, New Orleans Mayor Ray Nagin issued the city's first-ever mandatory evacuation order. He also declared that the Superdome, a stadium located on relatively high ground near downtown, would serve as a "shelter of last resort." Since this disaster was highly publicized, it was the subject of a huge number of studies [36–39]. (For a documentary movie on YouTube, see [40].)

The storm caused a great deal of damage, but the aftermath was catastrophic. First of all, it resulted in 1863 deaths. The Coast Guard rescued about 34, 000 people in New Orleans alone. However, the federal government was slow to respond. Specifically, the Federal Emergency Management Agency (FEMA) did not seem to have any sound plan of action, and they took days plan and execute operations.

Mistakes That Led to the Disaster

Hurricanes are part of New Orleans' history, and its vulnerability to flooding was well known. Even though another hurricane was anticipated, the city was not well-prepared either physically or socially for the catastrophe.

As concerns physical preparation, it is true that the U.S. Army Corps of Engineers built a system of levees and seawalls. Levees along the Mississippi River proved to be strong enough to defend the city from flooding. However, structures built near the city's east and west were much less well grounded. The Army Corps of Engineers acknowledged some responsibility, admitting that the levees failed due to the flawed and outdated engineering practices used to build them.

Here we mention just one example of the lack of social and financial preparation. FEMA failed to classify many areas of New Orleans as floodplains prior to Katrina. Residents who live in floodplains should have a flood insurance policy controlled by the National Flood Insurance Program. While residents saved some money by not paying for the insurance, they could not apply for compensation after the catastrophe.

The Rebuilding Process

In a nutshell, the rebuilding of New Orleans is still a work in progress, and some of the disaster's impacts will likely never be fully erased from the city. However, 90 percent of New Orleans' population has returned, and the Army Corps of Engineers built a new system of levees and floodwalls around the city.

Frederic Schwartz, an architect selected to participate in the rebuilding, describes the situation as one we may call an example of creative destruction, a concept we will discuss in Sect. 5.1.2:

> The planning of cities in the face of disaster (natural and political) must reach beyond the band-aid of short-term recovery. Disaster offers a unique opportunity to rethink the planning and politics of our metro-regional areas–it is a chance to redefine our cities and to reassert values of environmental care and social justice, of community building and especially of helping the poor with programs for quality, affordable, and sustainable housing [41].

Lessons Learned from Hurricane Katrina

Disasters are times of disruption, trauma, and even horror. But they are also moments when people can come together to help each other as equals. It is not surprising that people with greater numbers of formal and informal connections bounce back much better than the lonely, isolated ones.

We are slowly beginning to understand the need to focus on emergency preparedness well before disaster strikes. The elements of community resilience are both physical and social. Resilience requires the ability to repair infrastructure (buildings, roads, and levees), improve methods of predicting hazards, and provide rapid disaster response. Advice for building resilient communities tends, unfortunately, to be more normative than operative. Here is some operative advice: Cooperation among neighbors improves communication in times of emergency, participation in design processes increases feelings of belonging, and the inclusion of the most marginalized residents in a support system leaves fewer people out of a functioning community.

Somewhat paradoxically, disasters like Katrina and now many others prove that another world is possible. It is possible to make a transition to a world in which efficient responses to disasters may arise as a result of local people helping each other. A community can be resilient only if the individual people cooperate and show solidarity, meaning that even individuals must demonstrate resilience.

4.2.4 COVID-19: Rapidly Changing Perspectives

We write a book that we hope will be readable for years, not an article for a weekly magazine, and the "silver bullet" for handling the terrible pandemic is certainly not in our hand. Still, we cannot neglect the topic. Perspectives on how to handle a pandemic have changed as a result of COVID-19, reflecting a combination of both magical and wishful thinking with scientific and technological progress. To put it into a broader context, we should recognize that the pandemic will not disappear from one day to the next. Based on general epidemic principles, global herd immunity, seen as the solution in the beginning, looks unreachable. New variants could return us to a crisis state once again.

New problems generally cannot be solved by old means. The pandemic really is global, but local and national governments have mostly been the ones tasked with coping with the pandemic. Even non-cynical governments look toward the next elections, so their policies might be influenced by short-term political goals.

From the perspective of complex systems theory, it is clear that we need greater overarching, strategic thinking. Even the international system, which should have responded to the pandemic appropriately, should be repaired [42]. As nationalism is rising, it is not clear how countries will recognize the necessity to work together to reform global public health institutions.

The irresponsibility of Chinese and U.S. political leaders in the initial phase of the pandemic is undeniable. It was against all odds that the United States, the home of the world's leading scientific institutions, accounted for 25 percent of the world's COVID-19 cases and 20% of deaths from the disease during the first year of the pandemic, despite the fact that the United States constitutes just four percent of the global population.

Vaccine development, however, is certainly a success story. Pharmaceutical and biotechnology companies and governments cooperated to come out with efficient vaccines as rapidly as possible. Messenger RNA (mRNA) vaccines teach our cells to make a protein that will trigger an immune response inside our bodies. The science underpinning the vaccine has been known for decades, but it became the most important technique in the fight against COVID-19.

While the creation of vaccines was a real success, their distribution was much less so. Bilateral agreements to share the vaccine are better than nothing, but they are far from being optimal, and there are no global coordinators for doling out vaccine doses. Wealthy countries are reluctant to share the vaccine with less-well-off countries. China has exported more than 200 million doses of vaccines, but they have not necessarily been well-tested.

The most ambitious global enterprise is *COVID-19 Vaccines Global Access* (*COVAX*). COVAX is an international initiative, mostly founded by Western countries, with lofty goals regarding the distribution of vaccines to low-to-middle-income countries. COVAX promised 2 billion vaccine doses to help the world's neediest in 2021. It won't deliver even half that [43]. The World Health Organization (WHO) set a vaccination rate goal of 40 percent, but at the end of 2021, only about 9 percent of people on the African continent have been fully vaccinated. The lack of appropriate infrastructure, and of funding for training and deploying medical staff, have played a role.

Where Are We, and What Should We Do?

First, we have normative recommendations: ("you should do this and that"):

- Wealthy countries must invest in the next generation of COVID-19 vaccines, ones that are less expensive to manufacture, require no refrigeration, and can be given in a single dose by untrained personnel.
- The UN should create a "global health threats council." While this council would cooperate with the WHO, it should be a separate institution. The council should make countries accountable.
- The world must work together to build an enduring system for mitigating this pandemic and preventing the next one.

Second, we have the big question regarding how to implement these recommendations. It appears that already advocates of a "vaccine nationalist" approach and those who hold the globalist perspective of coping with pandemics have begun to fight [44]. Our cautious optimism suggests that some new global institutions will emerge to save the world.

The participants at a September 2020 workshop held at the Fondation Brocher in Geneva, Switzerland analyzed how to build resilience to pandemics. Their suggestion was that in addition to health systems, the economic, environmental, and governance systems should be integrated. Societal resilience is an emergent property of complex systems acting across different scales and levels of social organisation and governance [45].

4.3   Right to Repair, Fight to Repair

A few years ago, Zsuzsa's family washing machine broke down when it was just four years old. They spoke to three mechanics, but all of them said that the motor was seriously defective, that it needed to be replaced, and that it was better to buy a new washing machine right away. Zsuzsa could not believe her ears. With three children, four or five washes a week can't ruin the machine in such a short time! Would she have to buy a new machine every four years? The third mechanic scratched his head moodily, clearly not happy with this solution either. Finally, he gave a tip on what brand of washing machine to buy—and the machine has been working fine ever since. That's not to say that it's worth knowing the brands of less perishable machines: It's just that the experienced professional's tip worked. And frankly, it's a great satisfaction to see the confusion and anger of a skilled and experienced technician confirming that something is not right.

The Throwaway-repair Dilemma

This section is about the "right to repair movement" and the growing desire among consumers to fix products rather than throw them away. Although this seems like a good thing to do, companies have not always made it easy.

Recently, the daughter of Ontario Liberal Party member Michael Coteau dropped her smartphone, prompting outrage from Coteau that getting the manufacturer to repair the phone was more expensive than simply buying a new phone. It was he who introduced a "right to repair" bill for electronic devices in the Ontario provincial parliament in 2019 [46].

Decades ago, manufacturers provided full documentation for each of their products, with specifications for each component. Today, on the other hand, they have increasingly sophisticated ways of making it impossible, expensive, or at least difficult to repair broken products. Technology is advancing at lightning speed, but the aim is not durability. Instead, it is a precisely defined (preferably not too long) service life. Repair is a business loss for the manufacturer and a gain for the environment and the consumer: This is the repair dilemma. If I buy a blender and it breaks down unexpectedly, I don't want to pay 70–80% of the original price to have it repaired in a brand service center, where the machine will rest for weeks until the right part arrives. The part may even have a unique identification number, so if they replace it with the same one at a nearby independent repair shop, it still may not work. If you had a lot of time and a lot of money, you could sue the mechanic.

A recent survey showed that nearly 80 percent of European Union (EU) citizens would rather repair their appliances than replace them. Sill, significant proportion of generally "reasonable" people often waste in ways we may call "unreasonable." The majority of them explicitly believe that manufacturers should be legally obliged to make it easier to repair digital devices [47].

If you have been using the above-mentioned blender for, say, three years, you have to wonder whether it might not be a better decision to buy a new one instead of pursuing the complicated and expensive repair. If the repair shop quotes a fair price with a good time frame, we will get it repaired without a second thought. While this is favorable for the customer, and for the planet, the manufacturing company doesn't always agree: they see it as a lost profit. It defines its own business interest as selling as many products as possible in as short a time as possible.

Smartphones and laptops are much more important than blenders. Today, it is still the manufacturer's legal right to prohibit the installation of reconditioned parts obtained from a scrapper or installation and repair by a technician who is not authorized by the manufacturer.

However, this severely restricts the rights of owners who pay a hefty price for a piece of equipment. A replacement part is obviously cheaper than a new appliance, and an appliance that can be repaired will not be thrown away immediately. This means that, on the one hand, a lot of harmful waste does not accumulate and, on the other hand, a new item will not be produced using many more resources. Mechanics are very effective environmentalists, whether they know it or not.

But for me, as a customer, the company leaves me only the option of buying the item at a high price and then throwing it away, thereby increasing the amount of harmful waste. This is the dilemma of repair: What can be repaired will not only not be thrown away, but people will not buy a new one until they have to. This goes against the interests of the manufacturing companies.

The Genie Is Out of the Bottle

When we buy a product, we don't mentally prepare to throw it away after the first failure. If it breaks down, it still has to last. It has to be repaired. But here is a problem: Manufacturers should design repairable products, and the technical support of repair professionals and businesses is also very much needed. This is also the demand of the European organizations that have been calling for the right to repair on a common platform since September 2019 [47]. One of them is the Belgium-based ECOS, an international non-governmental organization that uses its network of experts to promote environmentally friendly technical standards and legislation. Another is the European Environmental Bureau, Europe's largest network of environmental citizens' organizations, which aims to ensure that consumers can repair their appliances more cheaply, without having to constantly buy new products, and that service companies can offer their customers repairs for different brands of equipment.

The iFixit website has the following slogan: *If you can't fix it, you don't own it.* Repair is better than recycling. Making our things last longer is both more efficient and more cost-effective than mining them for more raw materials. iFixit Europe, based in Germany, is a free, publicly editable online repair database (Free Repair Manual) with over 50, 000 manuals [48]. It is visited by tens of millions of people every month, and more and more of them are becoming empowered to fix their things. This is both saving people money and reducing mountains of electronic junk.

Recycling is cheap relief for the conscience of the wasteful consumer. It's easier to throw away a complicated object, which in the case of mobile phones contain precious metals, if we delude ourselves into believing that recycling will sort it out. It won't: Not even a fifth of e-waste can be recycled [49].

The campaign for the right to repair is spreading rapidly, with more than 40 organizations from over 16 European countries now working on the issue. Represented among these groups are community repair groups, environmental activists, social economy actors, independent repair advocates, and citizens ready to repair. They are people and organizations fighting to remove barriers to repair and extend the lifespans of products and accusing the manufacturers themselves of making their products increasingly difficult to repair.

Products must be designed not only to perform but also to last and be repairable when needed. That is why the European initiative aims to ensure that EU legislation sets minimum design requirements that ensure that key components can be easily dismantled and replaced [50]. The requirements also set out that spare parts and repair manuals must be made accessible, and the products are accompanied by a "Scoring System on Repairability" as part of the existing energy label for all energy-consuming products [51]. Incidentally, iFixit has been using a Repair Index for years, a scale from 1 to 10, which reflects not only the cost of repair but also its complexity.

The messages of the Right to Repair movement are addressed to state legislators. In order for a customer to repair their own items, the state must require large companies to make information about repairs public. But beyond manuals, standards, and guides, let us not forget that this is a movement that has as much to do with changing mindsets and attitudes as it does with repairing appliances.

Indeed, only consumers can change the manufacturers by bringing the issue to the attention of the relevant public institutions to change the regulations on manufacturers. But we cannot get ahead of ourselves: It will take several years before the right regulations are in place to repair all types of products.

Manufacturers are interested in quick profits, consumers in sustainable development. But beyond the movements, we also need to transform the information and repair infrastructure. Manufacturers are likely to argue that the availability of repair guides, source codes, and specifications will make it easier for hackers, extortionists, and cybercriminals to mess with digital devices. They are also likely to oppose providing access to intellectual property, which can introduce competition risk. But far less is said publicly by manufacturers about the benefits they would lose if customers did not take their devices to official brand repair shops.

The Revolution Has Begun

The right to repair has received increasing attention in recent years. Technologies are changing rapidly, as are manufacturers' attitudes to the right to repair. Every consumer has the right to buy something and, if it breaks down, to have it restored to its original condition by a repairer.

In the United States, Massachusetts is a pioneer in the movement, with a 2013 law requiring automakers to provide diagnostic and repair information to the public and independent repair shops for vehicles manufactured after 2015. And although the law only went into effect in Massachusetts, automakers have expanded their information provision to other U.S. states [52].

There are signs of action at the federal level, as well. During the summer of 2021, U.S. President Joe Biden signed an executive order asking the Federal Trade Commission to develop a uniform set of rules for the repair and supply of parts for farm equipment and machinery: The result will be that many farmers will be able to repair their own machinery [53]. In New York, the Fair Repair Act, which requires manufacturers to make diagnostic and repair information available for all products, has passed the state Senate [54].

In 2020, the European Parliament adopted the Right to Repair resolution with 395 votes in favour, 94 against, and 207 abstentions. So there is hope that mandatory labeling on the life expectancy and repairability of a product will be introduced. In March 2021, the EU put into force a regulation on the right to repair for TVs, dishwashers, hairdryers, and refrigerators. Manufacturers of these appliances are obliged to ensure that products can be repaired for 10 years. In addition, a label will be placed on the appliances indicating the expected lifetime of the product. They will be accompanied by repair instructions and will be manufactured in such a way that they can be dismantled and repaired using common tools.

In January 2021, France introduced a new repairability index for five categories of electronic devices, including smartphones and laptops. The index aims to inform customers about the repair options available for products before they buy them.

The new attitude will open up a new world for ordinary people, who will feel responsible in ways beyond saving money and protecting the environment. We will also see the fridge in the corner of the kitchen become a familiar object in the family, and the electric cooker will no longer need to be replaced every few years. Familiar objects have familiar motions, which adds to the comfort and coziness of the home space. My (Z) grandmother always handled the coffee grinder with special care because it cracked once—that was not a reason to throw it out but to change the relationship with it and perhaps be even more respectful of the treasured piece, which held up after its minor injury. We resonate very much with her perspective on a post-throwaway society.

Davids and Goliaths

Alongside—or against—the giant corporations that look like Goliaths, more and more "Davids" fighting for the right to repair are appearing in the public eye and going after the big corporations. Among them is 33-year old Louis Anthony Rossmann, a New York-based independent repair technician and right-to-repair activist who regularly speaks and testifies at right-to-repair hearings. In a 2018 investigative article for CBC News, it was revealed that Rossmann repairs Apple products much faster than the manufacturer's brand service [55]. He shares his repairs—and his thoughts on life and so much more—on his popular YouTube channel [56].

Another "David," Dutch Joanna van der Zanden, is not a mechanic but an art curator and a champion of repair culture, who one day wondered what society would be like if repair was central to our thinking and actions. She sees repair as a holistic solution and argues that we can repair relationships, biodiversity, systems—so let's repair them! [57]

Van der Zanden is credited with the launch of the Repair Café in 2009, which has since become a popular social movement thanks to the dedicated journalist Martine Postma. The curator is convinced that repair must precede recycling: This would mean that far fewer things are thrown away. But this requires a revitalization of the idea of repair and its reintegration into culture. She is also the author of the Repair Manifesto, which was launched with the Platform 21 team and artist-designer Cynthia Hathaway and has been translated into many languages [58]. Among other things, it emphasizes:

- Product designers should make their product repairable.
- Consumers should try to repair things before recycling.
- Repairing is a creative challenge.
- By fixing objects, you can contribute to their history.
- Repair is important even in good times! It has nothing to do with recession. It's about mentality, not money.

The Manifesto also highlights what is less discussed in policy platforms: the joy of repair—discovering the inner world of our broken things, the connection between parts and whole, and the hidden parts behind the operation until the moment of repair. The repair process is fun. It is so great to see how the repaired object comes to life— maybe it will take a different shape, maybe it will sound different, maybe it will get an extra accessory. And it comments on the joy of freedom and independence that a screwdriver set gives, knowing that you'll no longer be embarrassed by a loose dresser handle, and the expansion of that freedom when we share our proven repair techniques with our school-aged child, who will also be a notch freer from now on.

Before philosophy, action takes center-stage. Today, there are more than 2200 Repair Cafés worldwide [59]. These free meeting places are sites for communal repair, equipped with tools and materials, where visitors can repair their clothes, furniture, bikes, electrical appliances, and more, with the help of expert volunteers. They can also help each other and read books about DIY in a communal setting, over a cup of tea or coffee.

There are countless prominent Right to Repair movements and activists around the world: One more is worth mentioning because this one is a co-founder of Apple. Steve Wozniak, who saw the birth, growth, and transformation of the technological "Goliath" at close quarters in the 1970s talks today about Apple's changing philosophy and the need to recognise the right to repair [60].

Arguments are pouring in from both sides—manufacturers and the service community. It is perhaps too early to call it a dialogue. Because manufacturers can spend considerably more on lobbying, the mindset shifts that Repair Cafés have initiated on a small scale are so important. And they have started to introduce the repair approach to schools in pilot projects [61].

Repair First

For the millions of products sold each year, three very significant cash-generating activities are linked: accessories, servicing, and taxes. The last is a monopoly of the state, but the first two make any discourse on repair culture really about capitalism. Repair is the antithesis of the greedy capitalism that is rampant in the modern world. In the eyes of manufacturers, repair is an obstacle to future profit: The longer we use a device, the fewer devices they sell.

Murphy's well-known law applies to electronics too. What can go wrong, will go wrong. Even with the most carefully controlled manufacturing process, problems can still occur, and a person making a call can accidentally drop a device on the pavement. However, what can be repaired, must be repaired.

As the years go by, the feeling of vulnerability grows in the consumer. There are more and more services that require a cell phone and its various features. Think of the QR code reader, the camera, the GPS signal: More and more of our daily activities depend on them. The changing role of cell phones is particularly affecting those who have difficulty accessing services, for example because of where they live.

The enemy of Right to Repair is also planned obsolescence. As was discussed in Sect. 1.2.2, planned obsolescence, or premature obsolescence, is good for manufacturers because it encourages consumers to buy new products more often. However, premature obsolescence is caused not only by deliberate product failure but also by other planned processes. This is the case when spare parts are very slow to arrive at the service station, or where the cost of the spare parts adds up with the cost of the labor, and the customer is more likely to buy a new appliance when faced with the substantial sum involved. It could also be that a software update is delayed, or the repair shop is far away, or it is simply too inconvenient to have the repair done. In other words, the repair will be slow, expensive, and complicated.

And once again, we bury our heads in the sand using the excuse of recycling. We should remphasize: E-waste—which is growing by two million tons per year around the world—is made up of our discarded phones, kettles, sports equipment, toys, TVs, and other digital devices. The mass of e-waste generated in 2021 was bigger than the Great Wall of China. (We cannot deny that this a somewhat populist comparison.) This estimate was produced by the WEEE Forum, an international group of experts working to tackle the global problem of discarded electrical and electronic equipment, and they found that the value of e-waste is greater than the GDP of many countries [62]. As Rüdiger Kühr, director of the UN's Sustainable Cycles (SCYCLE) programme, put it, "A tonne of discarded mobile phones is richer in gold than a tonne of gold ore" [49].

The volume of e-waste is growing fast, with 53 million tons generated annually. The best way to reduce this is to repair appliances. It is in the interests of individuals and humanity to spend less money on new products and generate less waste.

Whatever Murphy's Law of Broken Things may say, if it can be fixed, it should be fixed.

4.4   Lessons Learned and Looking Forward

In this chapter we reviewed basic mechanisms for responding to disturbances, which can drive systems back to their original states. It looks Nature provides some compensatory mechanisms to promote adaptation to the continuously changing environment.

The observation, which became known as Le Chatelier's principle, speaks about the stability of a thermodynamic equilibrium state, but its spirit was soon extended to to biological and social systems too.

Homeostasis should be considered as a general repair mechanism that ensures the functional stability of living systems. Living systems are in continuous interaction with their environment through material, energetic, and informational flows.

As we learned from the somewhat forgotten field of cybernetics, the stable operation of biological and social systems is maintained by the balance of positive and negative feedback loops. It sounds like an oversimplification, but basically it is true: When the balance is violated, huge systemic failures, such as social inequality and climate change, emerge.

The concept of resilience emerged from the field of ecology, but its applicability and significance has escalated in light of natural and social disasters. A system is resilient if it continues to carry out its function in the face of adversity. We discussed the importance of being resilient at many levels of hierachical organization, from individual to buildings to small and large communities.

In the age of the transition from the throw-away society to repair society a fight has started at least two levels. At the *mental* level, we should relearn what was known and not more or less forgotten. Our first reaction in times of crisis should be about how to repair or save something. At the *legal* level, the right to repair movement has become stronger. Giant corporations have made some concessions, but the war is not yet over.

References

1. Henderson L (1970) Sociology 23. In: L.J. Henderson on the Social System, vol 28. University of Chicago Press, pp 73–74
2. Szenberg M, Ramrattan L, Gottesman AA (2006) Samuelsonian economics and the twenty-first century. Oxford University Press
3. Diamandis EP (2021) COVID-19 and the Le Chatelier's principle. Diagnosis 8(4):445–446. https://doi.org/10.1515/dx-2021-0022
4. Billman GE (2020) Homeostasis: the underappreciated and far too often ignored central organizing principle of physiology. Front Physiol 11. https://doi.org/10.3389/fphys.2020.00200
5. Wiener N (1948) Cybernetics or control and communication in the animal and the machine. MIT Press
6. Grossberg S (1980) How does a brain build a cognitive code? Psychol Rev 87:1–51. https://doi.org/10.1007/978-94-009-7758-7_1
7. Grossberg S (2013) Adaptive resonance theory: how a brain learns to consciously attend, learn, and recognize a changing world. Neural Netw 37(1–47). https://doi.org/10.1016/j.neunet.2012.09.017
8. Barabási A (2018) The formula: the universal laws of success. Little, Brown and Company
9. Piketty T (2013) Capital in the twenty-first century. Belknap Press
10. Chen D (2020) Capitalism's positive feedback loop is in need of a negative one. The American Prospect. https://prospect.org/essaycontest/daniel-chen/
11. The study of Earth as an integrated system. NASA Science Global Climate Change. Accessed 17 Apr 2022. https://climate.nasa.gov/nasa_science/science/
12. How feedback loops are making climate crisis worse (2020) Climate reality project. https://www.climaterealityproject.org/blog/how-feedback-loops-are-making-climate-crisis-worse
13. Holling CS (1973) Resilience and stability of ecological systems. Ann Rev Ecol Systematics 4:1–23. https://doi.org/10.1146/annurev.es.04.110173.000245
14. Natural Disasters 2019 (2020) Centre for research on the epidemiology of disasters. https://reliefweb.int/report/world/natural-disasters-2019
15. St. Louis Sage (2021) What was St. Louis' worst heat wave? St. Louis Magazine. https://www.stlmag.com/history/what-was-st-louis-worst-heat-wave/
16. 2019 Chennai Water Crisis. Wikipedia. Last updated 17 Apr 2022. https://en.wikipedia.org/wiki/2019_Chennai_water_crisis
17. Building Your Resilience. American Pyschological Association. Last updated 1 Feb 2020. https://www.apa.org/topics/resilience
18. Van Dam NT, van Vugt MK, Vago DR, Schmalzl L, Saron CD, Olendzki A, Meissner T, Lazar SW, Kerr CE, Gorchov J, Fox KCR, Field BA, Britton WB, Brefczynski-Lewis JA, Meyer DE (2017) Mind the hype: a critical evaluation and prescriptive agenda for research on mindfulness and meditation. Perspect Psychol Sci 13(1):36–61. https://doi.org/10.1177/1745691617709589
19. Serenity Prayer. Wikipedia. Last updated 8 Apr 2022. https://en.wikipedia.org/wiki/Serenity_Prayer

20. Resilience: build skills to endure hardship (2020) Mayo Clinic. https://www.mayoclinic.org/tests-procedures/resilience-training/in-depth/resilience/art-20046311
21. A Practical Guide to Climate-Resilient Buildings and Communities (2021) United Nations Environment Programme, Nairobi. https://www.unep.org/resources/practical-guide-climate-resilient-buildings
22. Cohen B (2011) Global ranking of top 10 resilient cities. Triple Pundit. https://www.triplepundit.com/story/2011/global-ranking-top-10-resilient-cities/76411
23. Érdi P (2019) Ranking: the unwritten rules of the social game we all play. Oxford University Press
24. Weeks K (2014) Which cities are the most resilient and the most vulnerable? Architect Mag. https://www.architectmagazine.com/technology/which-cities-are-the-most-resilient-and-the-most-vulnerable_o
25. Dubai named world's most resilient city by United Nations (2021) The National. https://www.thenationalnews.com/uae/government/2021/09/23/dubai-named-worlds-most-resilient-city-by-united-nations/
26. US President. Executive Order 14017 of February 24, 2021 on America's Supply Chains (2021) Federal Register 86(38):11849–11854. https://www.govinfo.gov/content/pkg/FR-2021-03-01/pdf/2021-04280.pdf
27. Remarks by President Biden at Signing of an Executive Order on Supply Chains (2021) White House Briefing Room. https://www.whitehouse.gov/briefing-room/speeches-remarks/2021/02/24/remarks-by-president-biden-at-signing-of-an-executive-order-on-supply-chains/
28. Brunnermeier M (2021) The resilient society. Endeavor Literary Press
29. Kotler S (2021) How to recover from burnout. Psychol Today. https://www.psychologytoday.com/us/blog/the-playing-field/202101/how-recover-burnout
30. Goldberg M (2021) Feeling burned out? These expert-approved strategies will help you recover. Oprah Daily. https://www.oprahdaily.com/life/a36801181/how-to-recover-from-burnout/
31. Matravers D (2018) To restore or not to restore? OpenLearn. https://www.open.edu/openlearn/history-the-arts/philosophy/restore-or-not-restore
32. Cameron C (2017) Reconstruction: changing attitudes. The UNESCO Courier. https://en.unesco.org/courier/july-september-2017/reconstruction-changing-attitudes
33. Policies Regarding CONSERVATION of World Heritage Properties. UNESCO World Heritage Centre. Accessed 17 Apr 2022. https://whc.unesco.org/en/compendium/109
34. Foster A (2010) Building conservation philosophy for masonry repair: Part 1—ethic. Struct Survey 28(2):91–107. https://doi.org/10.1108/02630801011044208
35. National Register of Historic Places. National Park Service. Last updated 13 Apr 2022. https://www.nps.gov/subjects/nationalregister/index.htm
36. Plough AL, Chandra A (2015) What Hurricane Katrina taught us about community resilience. RAND Blog. https://www.rand.org/blog/2015/09/what-hurricane-katrina-taught-us-about-community-resilience.html
37. Remes J (2015) Finding solidarity in disaster. The Atlantic. https://www.theatlantic.com/politics/archive/2015/09/hurricane-katrinas-lesson-in-civics/402961/
38. Santos N (2019) Fourteen years later, New Orleans is still trying to recover from Hurricane Katrina. Environmental and Energy Study Institute. https://www.eesi.org/articles/view/fourteen-years-later-new-orleans-is-still-trying-to-recover-from-hurricane-katrina
39. Hurricane Katrina. History. Last updated 9 Aug 2019. https://www.history.com/topics/natural-disasters-and-environment/hurricane-katrina
40. Katrina, The New Orleans nightmare: documentary on the devastation of Hurricane Katrina. https://www.youtube.com/watch?v=JEAedjLXw7Q
41. Schwartz F (2007) New Orleans now: design and planning after the storm. In: Mateo JL (ed) Natural metaphor: an anthology of essays on architecture and nature. ACTAR
42. Brilliant L, Danzig L, Oppenheimer K, Mondal A, Bright R, Lipkin WI (2021) The forever virus. A strategy for the long fight against COVID-19. Foreign Affairs. https://www.foreignaffairs.com/articles/united-states/2021-06-08/coronavirus-strategy-forever-virus

43. Taylor A (2021) Covax promised 2 billion vaccine doses to help the world's neediest in 2021. It won't deliver even half that. Washington Post. https://www.washingtonpost.com/world/2021/12/10/covax-doses-delivered/
44. Zhou YR (2021) Vaccine nationalism: contested relationships between COVID-19 and globalization. Globalizations 19(3):450–465. https://doi.org/10.1080/14747731.2021.1963202
45. Wernil D, Clausin M, Antulov-Fantulin N, Berezowski J, Nikola Biller-Andorno N, Blanchet K, Böttcher L, Burton-Jeangros C, Escher G, Flahault A, Fukuda K, Helbing D, Jørgensen PS, Kaspiarovich Y, Krishnakumar J, Lawrence RJ, Lee K, Léger A, Levrat N, Martischang R, Morel CM, Pittet D, Stauffer M, Tediosi F, Vanackere F, Vassalli JD, Wolff G, Young O (2021) Building a multisystemic understanding of societal resilience to the COVID-19 pandemic. BMJ Glob Health 6:e006794. https://gh.bmj.com/content/6/7/e006794
46. https://www.youtube.com/watch?v=_XneTBhRPYk
47. Sieg K (2021) The EU is giving citizens the "right to repair" electronics—here's what that could mean for the world. TED. https://ideas.ted.com/how-right-to-repair-legislation-can-reduce-waste/
48. Repair guides for every thing, written by everyone. iFixit: the free repair manual. Accessed 17 Apr 2022. https://www.ifixit.com/
49. Gill V (2021) Waste electronics will weigh more than the Great Wall of China. BBC News. https://www.bbc.com/news/science-environment-58885143
50. Right to Repair Europe. Accessed 17 Apr 2022. www.repair.eu
51. What we want. Right to repair Europe. Accessed 17 Apr 2022. https://repair.eu/what-we-want
52. Robertson A (2020) Massachusetts passes 'right to repair' law to open up car data. The Verge. https://www.theverge.com/2020/11/4/21549129/massachusetts-right-to-repair-question-1-wireless-car-data-passes
53. Sink J (2021) Biden sets up tech showdown with 'Right-to-Repair' rules for FTC. Bloomberg. https://www.bloomberg.com/news/articles/2021-07-06/biden-wants-farmers-to-have-right-to-repair-own-equipment-kqs66nov
54. Godwin C (2021) Right to repair movement gains power in US and Europe. BBC News. https://www.bbc.com/news/technology-57744091
55. Canada gets closer to a right to repair law (2019) CBC. https://www.cbc.ca/news/science/what-on-earth-newsletter-right-to-repair-styrofoam-1.5037697
56. https://www.youtube.com/playlist?list=PLkVbIsAWN2lsx11ydFwiNZdQiZtOyeUV-
57. van der Zanden J. In repair: towards a post-throwaway society. Throwing Snowballs. Accessed 17 Apr 2022. http://www.throwingsnowballs.nl/index.php/writing/
58. Homepage. Platform 21. Accessed 17 Apr 2022. www.platform21.nl
59. About). Repair Café. Accessed 2022, April 2017. https://repaircafe.org/en/about/
60. Apple founder Steve Wozniak backs right-to-repair movement (2021) BBC News. https://www.bbc.com/news/technology-57763037
61. New starter kit: repair Café in the classroom (2019) Repair Café. https://repaircafe.org/en/new-starter-kit-repair-cafe-in-the-classroom/
62. 57.4 Mt e-waste expected in 2021 will outweigh China's great wall (2021) Recycling Mag. https://www.recycling-magazine.com/2021/10/14/57-4m-t-e-waste-expected-in-2021-will-outweigh-chinas-great-wall/

Chapter 5
The Pathways Toward the New Normal

Abstract The chapter studies the general mechanism for relocating from states considered normal (the "old normal") to a "new normal" in cases when a perturbation exceeds a certain threshold that prevents restoration to the original stable state. Biologists and social scientists have generalized theories from physics that describe transitions between stable phases, or equilibrium states, to biological and social phenomena. The lesson we take from this is that, in the overwhelming majority of cases in a dynamic world, impaired states cannot be restored to their origin. So, systems must often find a "new normal" behavior or state. Many natural and socio-technical systems operate close to critical points, tipping points, and places where small changes can have dramatic consequences. Humanity might be close to a negative tipping point. Nevertheless, we would like to believe that the transition from the throw-away society to the repair society has started.

5.1 When to Attempt Repair and When to Let Go

When we have an impaired system, we have to decide whether we should let it go or attempt to repair it. We let go of an old car when the cost-benefit analysis shows that it is more effective to buy a new one. If you are not happy with your language instructor because she teaches too many grammatical rules and not enough conversational skills, you may give her a second chance by making clearer your learning goals. But if something is inherently different from what you were hoping, it cannot be repaired. As the slogan says, a leopard does not change its spots. Otherwise speaking, "Fish gotta swim, birds gotta fly."

There is a general mechanism for relocating from certain states that were considered normal (the "old normal") to a "new normal" in cases when a perturbation exceeds a certain threshold that prevents restoration to the original stable state, as Fig. 5.1 shows.

We like to think that something or somebody shows normality. But everything and everybody has conditions and circumstances under which things change. Water is generally considered "normal" when it is in a liquid state. Changing the environmental conditions and reducing the temperature below one threshold, or increasing above to another creates a "new normal"—solid ice or vapor. The fundamental states

Fig. 5.1 If a large
perturbation occurs, the
system reaches a new
equilibrium state

of matter are solid, liquid, and gas, and the process of *phase transition* takes a physical system from one phase to another. A very special state is called a critical state or *critical point*. The liquid–vapor critical point is defined as the end point of the pressure–temperature curve that determines conditions under which a liquid and its vapor can coexist.

It is interesting to see that biologists and social scientists have generalized theories from physics that describe transitions between stable phases, or equilibrium states, to biological and social phenomena too. The lesson we take from this is that, in the overwhelming majority of cases in a dynamic world, disturbed or impaired states cannot be restored to their origin. This means that the system must find a "new normal" behavior or state, generally through a "jump." Small changes to systems can lead to sudden, drastic changes in their behavior, which in some cases appear as if they have emerged from the blue. These phenomena are associated with catastrophes, and *catastrophe theory* was first oversold and then underestimated as a mathematical framework for describing jumps between the old and the new normal.

5.1.1 About the Rise and Fall of Catastrophe Theory

René Thom (1923–2002), a French mathematician, who earned the Fields Medal (one of the most prestigious prizes for mathematicians) at a young age, was the first to classify these possible sudden jumps as "catastrophes" [1].

Thom applied strict mathematical theory to biological morphogenesis. As the theory became popular, it was adopted (sometimes superficially) to explain such social phenomena as aggression, economic growth, sudden price changes, stock market crashes, arms races, prison riots, and national war policy.

The basic properties of these systems include (i) bistability, (ii) discontinuity (catastrophe), and (iii) hysteresis. Bistability is associated with two states (say a "low price" regime and a "high price" one, as we will soon see in the next example). Discontinuity reflects the principle that "small changes in the causes may imply dramatic changes in the behavior." When a system responds to a changing environmental stimulus with some time delay, we have two paths (one for the case when the stimulus increases, and another one when it decreases) that don't coincide. This phenomenon called is hysteresis.

A Catastrophe Theory-Based Oil Price Model

An application of catastrophe theory using hypothetical data is that of a toy model of oil prices. The tacit assumption is that oil prices have either low or high value—they are two separate regimes. Occasionally, small changes in the circumstances create jumps from one regime to the other. The model defines two control parameters, and Fig. 5.2 illustrates the "cusp" or catastrophe that results with these parameters. The caption of the figure describes the possible scenarios in which a jump may occur. The whole modeling procedure is intuitive rather than technical.

The lesson learned is that while models may just be metaphors, we should see some patterns that describe the transition to the new normal. Around a critical point, small changes cause dramatic effects. A return to the original normal is not impossible, but it would happen on a different path. Catastrophe theory lost its popularity, partly because of superficial applications. However, its rise and fall is a colorful story in the history of science [2, 3]. Catastrophe theory became the victim of its own success and of the ambition of its pioneers (in addition to Thom, Christopher Zeeman (1925–2016), a British mathematician popularized both the theory and its application [4]). Zahler and Susmann sharply criticized catastrophe theory and most of its applications [5]. They claimed that such modeling efforts should be restricted to science and engineering and have almost nothing to do with biology and social sciences. While applications of catastrophe theory largely disappeared, the celebrated mathematician Vladimir Arnold (1937–2010) contributed to the deepening mathematical foundations of the theory [6].

The emotional attitudes underpinning the heated debate were certainly related to the methodological discrepancies between natural and social science. However, the attack was somewhat misdirected. First, Thom and Zeeman trained and worked as mathematicians. Second, while it was true that some applications were over-dimensionalized or unjustified, the attack weakened the general position of those who tried to use mathematical models in social sciences.

Next, we will turn to a general organizational principle found both in natural and social systems, called *self-organization*. It plays a role in *both* restoring stable states through self-repair mechanisms and forming new states when old ones prove unstable.

5.1.2 Self-organization: Transition to the New Normal

The Concept

Self-organization is a mechanism for maintaining and forming ordered behavior that occurs in both nature and society [7]. Macroscopic spontaneous order emerges from *local* interactions among parts of an initially disordered system. There are a huge number of examples—from laser physics to the spatio-temporal patterns of chemical and biological systems to the formation of communities, cities, and business cycles— that demonstrate this principle. From the perspective of this book, self-organization

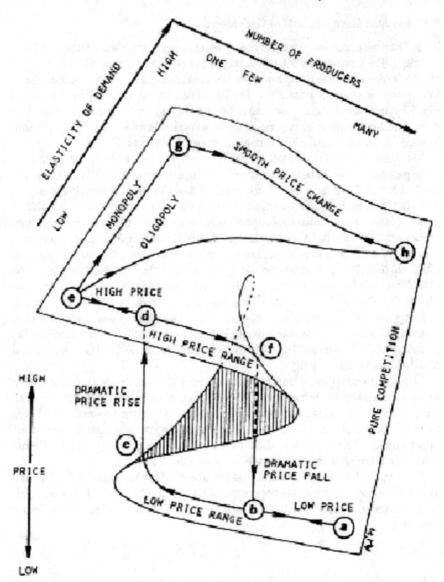

Fig. 5.2 The latitude and longitude in this case represent the elasticity of demand and level of competition in the crude oil market. The height of the landscape represents the price of oil. The model illustrates situations involving monopoly, oligopoly, and pure competition. The folded nature of the landscape surface suggests the existence of conditions supporting high and low price ranges. Paths such as $(a \rightarrow b \rightarrow c \rightarrow d \rightarrow e)$ on the landscape surface illustrate how decreasing competition can lead to sudden increases in price. Paths such as $(e \rightarrow d \rightarrow f \rightarrow b \rightarrow a)$ reflect sudden price declines due to increasing competition as new suppliers enter the market place. Increasing elasticity of demand can also lead to gradual changes in price (paths $e \rightarrow h$ and $e \rightarrow g$) under appropriate conditions. Adapted from Woodcock original figure republished in [2]

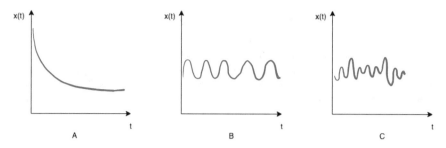

Fig. 5.3 Temporal patterns: **a** convergence to equilibrium; **b** oscillatory; **c** irregular (chaotic). Bifurcation between states may represent transitions to new normal states

may describe either self-repair as a response to external perturbations or the impetus for a transition from one state to another. Self-repair happens when the initial state is stable, meaning the original state can be restored after a disturbance. However, when the system loses its stability, it bifurcates to a *new* state. Self-organization often leads to the phenomenon of *pattern formation*. Patterns are ordered arrangements of constituent elements (e.g., molecules in chemical systems, species in ecological systems, stock prices, people as voters, and many others). There are temporal, spatial, and spatiotemporal patterns. Pattern formation is a process where an old, generally simple, pattern is destabilized and a more complex pattern associated with the "new normal" emerges. A branch of mathematics related to the study of nonlinear dynamical systems, called *bifurcation theory*, explains the conditions under which transitions occur and describes specific patterns [8]. Here, we give some illustrations of the different patterns and the transitions among them.

Temporal Patterns

Figure 5.3 illustrates three basic temporal patterns. Plot *a* embodies the convergence to equilibrium. Please note that unbounded growth processes in Chap. 3 lead to infinity, whereas here the process leads to a finite equilibrium point.

While the thermodynamic arrow of time suggests that processes tend to equilibrium in open systems, other temporal patterns such as regular oscillatory and irregular chaotic behavior may occur in physical, chemical, biological, and social systems. A non-damping pendulum and a resting heartbeat are the most characteristic illustrations, but there are many examples in the applied science literature, from oscillatory chemical reactions to biological clocks to certain electrical circuits to business cycle models. Plot *b* shows sustained oscillation, ideally with constant amplitude and frequency. Plot *c* visualizes an irregular behavior related to chaotic processes. Even very simple nonlinear dynamical systems can exhibit unpredictable behavior, which seems to be random, even though the generating algorithm is purely deterministic. This seemingly unpredictable behavior is often called chaos. We would like to ask the Reader to participate in an experiment. Today (January 10th, 2022) Google provided "About 331,000 results (1.07 s)" to the prompt "chaos is the new normal." How many responses do you see now?

Recent studies have posed the question whether COVID-19 transmission dynamics show chaotic patterns. The pandemic is developing simultaneously at multiple temporal and spatial scales, ranging from the individual to global. Particular disease patterns result from major self-organization mechanisms within the system. Chaotic behavior is a common phenomenon in complex systems, and most likely the unpredictability of the pandemic in different countries is a fundamental property of this dynamic system that demonstrates chaotic behavior [9].

Spatial Patterns

Computer scientists remember Alan Turing (1912–1954) for his fundamental contribution around 1936 to the theory of computability. Cognitive scientists celebrate his 1950 paper on "Computing machinery and intelligence." For biologists (mostly mathematical biologists), Turing's main achievement is the 1952 Royal Society paper "On the chemical basis of morphogenesis." Of course, as everybody knows, he played a major role in breaking secret German codes during World War II. His advances in different fields share a commonality: Algorithms that describe local interactions among components can produce ordered structures.

Turing wanted to show the possibility of spatially heterogeneous (but temporally stationary) stable structures emerging from a perturbation of (spatially) homogeneous structures. With a little overstatement, Turing wanted to ask the question of how something is generated from nothing. To be a little more specific: What is the source of spatial order? One possible philosophical answer is that order is created by some external agent, and the other is that it is produced by the interaction of constituents.

Turing's paper described how reaction-diffusion models found in mathematical chemistry might generate patterned structure during animal development in the embryonic state. Animals start as a single cell that divides many times to create a full-size individual. During the early stages, the small ball of cells is completely uniform, or homogeneous, but out of this develops the dramatic patterns of a zebra, leopard, giraffe, butterfly or angelfish. Turing showed that starting from a uniform ball of cells, spatially heterogeneous but static patterns (such as the stripes of a zebra) can be generated. Turing created a mathematical model that, under different conditions, led to a diversity of wonderful patterns that animals might have. Experimental demonstration and extensive studies of Turing structures started almost a half century later. Some basic patterns are shown in Fig. 5.4.

If the Reader is interested in "How the Leopard Gets Its Spots," a celebrated mathematical biologist wrote an excellent paper without including a single equation [11]. Turing's general question—using the modern terminology, how spatial order is formed from uniform but random structures without external instruction—and the principle of self organization reached the social sciences too.

Urban segregation is a celebrated example of such spatial patterns. Thomas Schelling (1921–2016) published a very influential paper in 1971 with the title "Dynamic Models of Segregation" [12]. This paradigmatic paper demonstrates how local rules ("micromotives," in Schelling's terminology) create global ordered social structures ("macrobehaviors"). Technically, it is a model involving *cellular automata* in one- or two-dimensional space. Each player (or agent) sits on a grid point sur-

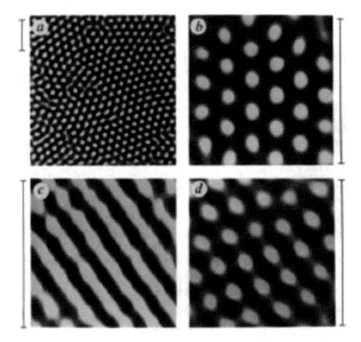

Fig. 5.4   Different spatial patterns (called Turing structures), such as blobs and stripes, occur in the same system for different parameter values. Adapted from [10]

Fig. 5.5   Randomly mixed and highly segregated patterns

rounded by others. (You may imagine that the points are, for example, people living in a house encircled by their neighbors). The players have a predominant parameter, which, in this case, is a visible color that correlates to the race of the players, since the model describes *racial segregation*. Figure 5.5 shows a randomly mixed and highly segregated pattern. Segregation patterns are distant relatives of Turing structures. They are relatives, since the generating rules are local but send signals that reinforce and inhibit each other. They are only distantly related, however, since the interactions are obviously not direct analogues of diffusion and reaction.

Residential segregation is ubiquitous in many countries. There is a very famous image captured by the photographer Tuca Vieira, which illustrates segregation in São Paulo, Brazil. Parasiopólis, a favela, is adjacent to the affluent district of Morumbi

[13]. It is a great example of how just a few streets can separate extreme poverty and extreme wealth. Brazil has extreme income inequality, and half of the state's population earns only 15% of the total national income. About 10% of people live in urban favelas [14]. It appears as though positive feedback increases inequality.

What policies could and should be implemented to break the vicious cycle of inequality? According to one source, the following policies are needed [15]:

- Promote housing affordability: Land-use regulations that are not too restrictive to new developments and suitable social housing systems that does not lead to a concentration of disadvantage
- Promote individual opportunities in the long term: Adequate provision of high-quality education and training available at the metropolitan scale
- Promote access to transport and jobs: Transport policies that connect employment and residential locations where needed
- Prevent isolation: Public spaces promoting interactions and livable communities

As it often happens, these recommendations are general, but what really matters is how the suggestions are implemented locally.

Urban segregation is a well-studied, popular example of social self-organization, but there are other examples of the phenomenon.

Social Self-organization: Some More Examples

While there are some hermits, people live in a society, and there are some principles that keep our society together. Nobody forced us to adopt these principles, but we have self-organized.

Human society evolved from small-scale, relatively egalitarian tribes. The structure of society was once relatively simple—there were more or less isolated small groups. Competition and cooperation among these groups led to the complex social entities of advanced industrialized nations. Since resources are limited, tribal groupings had to compete and act selfishly for the benefit of group members. However, historians and anthropologists have noted that during periods of intensive competition, like wartime, tribal groups also tended towards cooperative practices. Increased cooperation among group members induces firmer social cohesion, drives technological progress (including in military and organizational applications), and results in population expansion. Due to cognitive bounds that limit the number of social relationships any single person can maintain (the Dunbar number is about 150, as was mentioned earlier), evolutionary mechanisms have promoted the demarcation of social groups along cultural, linguistic, religious, and other lines. As populations have grown, humans have constructed ever-larger social hierarchies that now encompass societies of quite literally billions of people.

Human society now faces big challenges, as we have already discussed, like natural and social disasters, huge socio-economic inequalities, climate catastrophes, pandemics, and political instabilities. Complex natural and socio-technical systems often operate close to critical points. Under these conditions, a small trigger is enough to cause a major crisis. For example, maybe just a single stone is thrown by one demonstrator, but then a riot, a huge collective phenomenon, emerges. Once ignited, it cannot be controlled. There is a strong qualitative difference in whether the collective behavior of a mass of people is "peaceful" or "violent." There is a fire, or there is no fire.

Schelling studied a variety of social phenomena in which individual decisions were influenced by the behaviors of the others in the group. We may cross the street when the light is red if others around us do the same; we buy goods similar to those our neighbors buy; and so on. Schelling intuitively foresaw that there are situations with basically two outcomes (i.e., they show bistability)—two macrobehaviors emerge from the micromotivations of the individuals. Either all people cross the street (since the positive feedback of observing others cross amplifies the motivation), or everybody decides to wait. Either everybody starts to applaud after a performance, or weak applause decays very rapidly. Which of the two possibilities will be realized is determined by the number of people involved. If it is larger than a *critical mass*, then everybody (well, almost everybody) will join in.

Let's now discuss a less dramatic situation. The collective behavior of pedestrian crowds is a textbook example of how the interaction of a large number of "microscopic" elements that can lead to the formation of macroscopic patterns. Recalling Nobel-prize-winning physicist Murray Gell-Mann's (1929–2019) famous quote— "Imagine how hard physics would be if electrons could think"—it is somewhat surprising that a relatively simple model of individual human activities has the power to explain and predict complex collective behaviors. Dirk Helbing, a leading computational social scientist at ETH Zurich, made a few basic assumptions about individuals in his model that describes the movement of crowds [16]:

1. They show goal-seeking behavior (say, to walk or run from terminal A to terminal B)
2. They cannot walk through solid obstacles
3. There is a distance they feel comfortable putting between themselves and others (different cultures perceive this distance differently)
4. They want to be close to those that are like them.

The Social Force Model for Pedestrian Motion assumes that people are subject to three forces [16]. The first is an attractive force toward an intended goal. Second, there is a repellent or attractive force between any two people. Third, there is a repellent force between any people and obstacles.

Based on this small set of assumptions regarding the behavior of individuals, the model describes population behavior surprisingly well. Among other things, it explains the emergence of lanes of people walking or running in uniform direction. Such self-organized patterns of motion demonstrate that efficient and "intelligent" collective dynamics can be based on simple, local interactions. Under extreme crowd densities, however, coordination may break down, giving rise to mob- or stampede-like behavior. *Panic* is a collective behavior with potentially tragic outcomes. There is some possibility of preventing the breakout of panic by optimizing the design of pedestrian facilities, in particular for evacuation situations. The simulation of collective phenomena in the framework of self-driven, multi-agent systems is a useful method for identifying strategies to diffuse or escape panic [17].

From Consensus to Polarization and Back

Interactions among people in smaller or larger groups may cause changes in their opinions about different issues, and this may lead to *consensus*, *fragmentation*, or *polarization*. Consensus means that all players share the same opinion, fragmentation occurs when several opinions emerge, and polarization is a special case of fragmentation in which two groups of individuals with different opinions arise.

Opinion dynamics describe the process in which individual opinions change due to the interactions of players. Players update their opinions on the same issue based on some "rule" that expresses how others' opinions influence yours. We know that media influences our opinions, so changes in our opinions may depend both on local and global effects.

The concept of *bounded confidence* reflects the assumption that players with starkly different opinions do not influence each other [18]. Only when there is relative congruence between the opinions of players will they have an effect on one another.

It is not easy to maintain diversity of opinions in a community. In any group, there exists some norm enforcement by the majority, and there are some biases toward conformity that help generate consensus. On the other hand, aversion toward other groups may lead to strong inter-group differences, and in its extreme form, polarization. Too much consensus can be suspicious. The author(s) still remember Hungarian elections that were won with supposedly 98.4% of the vote but that did not necessarily reflect a deep consensus. Recently we see something very different, but we don't have clear answers as to why polarization has become a predominant phenomenon in many places in recent years, even though we can identify recurring patterns. First, through a variety of different mechanisms, decisively polarizing leaders attain power. Initially small differences in political power are amplified by the media. One important element is the manipulation of judicial systems, which leads to distrust from opposing sides. It looks as though polarization has become particularly strong in the United States. Once again: Will polarization be the new normal, or can it be tamed somehow?

Thomas Carothers and Andrew O'Donohue from the Carnegie Endowment for International Peace (a nonpartisan international affairs think tank) edited a book that investigates the question of why polarization has arisen in countries ranging from Brazil, India, and Turkey to the United Kingdom and United States [19]. Most importantly, once societies have become deeply polarized, would it constitute a "new normal," or are there any healing mechanisms that can generate a healthier distribution of divisions? Carothers and O'Donoughe seem to be optimistic that (i) *institutional reforms*, (ii) *legal or judicial actions*, and (iii) *political leadership* may play a crucial role in de-escalating partisan divides. It is too early to see whether or not these mechanisms will be implemented to successfully fight polarization. To preserve the (cautiously) optimistic tone of this book, we may argue that crisis does not necessarily mean the end of the world, as we discuss in the next paragraph.

From Crisis to Opportunity

Here is our consolation: While we lose some part of the world with which we are familiar, even forced change may bring the possibility of a new beginning. As we mentioned in the previous chapter when we discussed the rebuilding process after Hurricane Katrina, a crisis can be translated into the opportunity to start again. Maybe on the same path or maybe on a different one but, in any case, with more experience. Self-organization might be the new normal of global and local business crises. Such crises in Japan are known as "opportunity for improvement" (Kaizen). Kaizen is based on four principles [20]:

1. Every process can be improved
2. Defects and process failures are usually the results of imperfect processes, not people
3. Every person in the organization must have a role in improvement
4. Small changes can have a significant impact.

Self-organized systems are those that can evolve and improve behavior and structure to better adapt to environmental conditions. The decentralization of decision-making might be more efficient than centralized processes (at least when you have time to make a decision but perhaps not when you are trying to avoid being eaten by a hungry lion).

Creative Destruction

Joseph Schumpeter (1883–1950), the father of the concept of creative destruction, suggested that capitalism, even after crisis, creates new products and markets and new organizational principles that sweep away the old. Henry Ford's assembly line is an historical example of creative destruction and led to a revolution in the automobile manufacturing industry. The emergence of Netflix is another famous example of creative destruction. Its emergence led to the total disappearance of the video rental industry. As data show: "As of August 2021, Gen Zers were the most likely group in the U.S. to have a current Netflix subscription, with 78% saying that they subscribed to the service, compared to just 49% of Baby Boomers. Meanwhile, Gen Xers were the most likely to subscribe in the past, but not anymore" [21].

Video, and later DVD, rental shops used to be very popular in the United States and in Japan. The first step in the creative destruction of that industry occurred in 1997, when Netflix started an online rental service. People were able to mail-order DVDs to their houses. First, Netflix served those who did not have a video rental store nearby. The water became hotter when Netflix launched their subscriber-based business model in 1999. From 2007 on, there were some complicated steps toward what is now an unlimited streaming service. But the "water boiled to vapor" and resulted in creative destruction in the movie and TV industry, and the real destruction of the company Blockbuster (the big loser of technological progress). It has been documented that the leaders of Blockbuster missed the opportunity to negotiate with Netflix when the company was hardly even a rising star [22].

The statement is that growth comes from innovation, leading to new goods and increasing productivity. The price is the destruction of existing jobs and firms. But after the pandemic, or as we start to see now, between waves of the pandemic, the stakes of creative destruction will be much higher than simply trying to find innovative products. We might expect an adaptive recovery and accept that our life

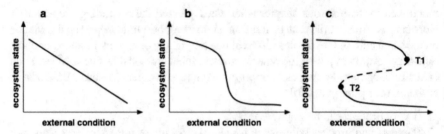

Fig. 5.6 Possible system changes [24]. The figure shows linear and nonlinear change in the response of a system in an equilibrium state to changing conditions, like water quality responding to changes in nutrient loading. The first type of response **a** is linear, incremental, and smooth following the increasing conditions, and may be reversible along the same trajectory if the driver of the conditions is decreased or restored to previous levels. The second type **b** shows threshold-dependent, nonlinear behavior, where the system moves from one state to another. In reality, this may also be quite easy to reverse. The third type **c** also shows threshold-dependent behavior, but a special type known as a critical transition or catastrophic fold bifurcation where the upper line state reaches a tipping point at T2, before dropping to the lower line state. As the names suggest, these are threshold-dependent changes associated with abrupt, surprising, and irreversible shifts in the form and function of a system. The term "regime shift" is often used to describe this kind of major shift. Fold bifurcation is the mathematical formulation whereby there are actually three equilibrium states, but where one is extremely unstable (the dotted line). Systems that come close to this unstable state rapidly move to one of the other stable states as shown by the arrows. There are two ecosystem states for a wide range of conditions between the two threshold positions T1 and T2, where moving back from the lower line state to the upper line state means changing the conditions at least as far as T1

will not be the same at the individual level (when we will shake hands quietly?), in communities (what will be the new order in gyms?), and at the global social scale (what will happen to conflict between nation states and global organizations?).

Thresholds and Tipping Points

The concept of *tipping points* was popularized by Malcolm Gladwell's bestseller *Tipping Point* [23]. Again, as the theory goes, the state of a system changes slowly for a while, but once you reach the edge of the abyss, it might be a good idea to stop, if you can. If the path exceeds a tipping point, a "new normal" may emerge. Figure 5.6 illustrates three ideal systems' answers for changes in the conditions.

5.1.3   Big Cultural Changes: They Started Yesterday, and Maybe It Is Not Too Late

Our big goal with this book is to promote the transition from the throw-away society to the repair society, since we belong to that camp that feels that humanity is close to a negative tipping point. People in this camp would like to see the repair society become the new normal. Speaking again in normative language, society needs transformational changes that alter the norms, values, and assumptions under which individuals and institutions function. We need something similar to the paradigm change in science and technological innovation [25, 26].

Where are we now? We are in the initial phase of the transition between the throwway society and the repair society. We have slightly modified a comparative table composed by Andy Ma, an industrial designer [27]:

Throw-way society	Repair society
Replace the old with the new	Repair not replace
Believe that new is better than old	Believe that old is better than new
Breakage means disability	Breakage is part of history
Make profit by launching new model	Make profit by providing repair service
Designed for replacement	Designed for easy repair
Broken for replacement	Broken for starting a new life
Emotionally detached to objects	Emotional bonds with objects
Planned obsolescence	Planned reuse
One-night-stand relationship	Long-term relationship
Similar	Unique
Consume	Save

Our specific question is whether or not we will see a new global community of "repairists" from the distributed clusters of local communities that have new moral principles and urges to resist over-consumption. We write this book since we see possible pathways, and we would like to activate Readers to take part in the transition. We don't state that the transition will necessarily happen—only that similar cultural changes have happened in the past, and there are possible ways in which today's society can also succeed.

While the study of cultural change has mostly been the field of historians and political scientists, social psychology and complex systems theory now look competent to address the problem. The cultural evolutionary approach can be useful for understanding norm development.

Mechanisms Behind the Tipping Point

There are some distributed seeds of repairist communities. Ideas and attitudes propagate like viruses or maybe wildfires. Complex systems theory offers models for both the *epidemic-like* propagation of ideas and propagation based on an analogy to *forest fire*. To block an epidemic, societies must reduce the efficiency of the interaction between the infected and the susceptible. If you want ideas to propagate through the whole community, you should increase the efficiency of their spread. We need to explain to people that the human society needs change in its perspective.

In addition to the gradual but threshold-dependent transitions, another possible mechanism is *punctuated equilibrium*. The concept of punctuated equilibrium, which can be contrasted with the gradual hypothesis, is attributable to Eldredge and Gould [28]. Connie Gersick adopted the idea to challenge the traditional gradualist understanding of cultural change. Gersick postulated that cultural change instead occurs through a punctuated equilibrium pattern, alternating between long periods of little change and "brief periods of revolutionary upheaval" [29, 30]. Gersick introduced the concept of *deep structure*, arguing that at the heart of this paradigm is the idea that "complex systems are held together by a highly durable underlying order, or deep structure." The deep structure is what confines change during equilibrium periods to variations on an enduring theme—and it is also what "disassembles, reconfigures, and enforces wholesale transformation during revolutionary punctuations."

We might see this as an explanation of how to transform the throw-away society. The elements of our current society's deep structure are the beliefs that new is better than old, consumption is value, and there is no need to have emotional relationships with objects. If you feel that one of these three ingredients is not necessarily true, you are already transitioning toward living in a repair society.

With the COVID-19 adventure, we're all suddenly aware of how quickly our everyday lives and attitudes might change. This is a chance to bring exciting new ideas into circulation, such as that of the repair society.

5.1.4   Release or Repair: Guidelines for Decision Making

In a repair society, we have a different attitude toward our relationships. When will we be better off: when we fix an object or when we throw it away (more subtly, let it go)? The answer is not always clear and easy. Something similar is true for relationships, although it is undoubtedly much more difficult to "calculate" and compare the possible outcomes. There are some general principles that may help to make a decision to let go or to fix easier.

Let Go

We tend to let go when the expense of a repair exceeds the price of a new product. Repairing a garment bought second-hand for pennies can sometimes cost more than buying a brand new one.

As we started out discussing at the beginning of the chapter, it's worth giving a language teacher or personal trainer a second chance: If you indicate that you want more emphasis on grammar or less on cardio exercise, and change happens, it's not worth switching. But if you make five requests and none of them are taken into account, this may not be the person you need.

In friendships, if effort is asymmetric—if the same person always initiates meetings—the other person might realize and correct the situation before the relationship falls apart. But if there is no change after repeated signals, it is worth acknowledging that the friendship is over and it is worth letting go.

The stakes are very high in marriages, especially if there are children involved. The question of letting go is rightly raised when trust has been severely shaken. In her book *Moral Repair*, Margaret Walker writes that trust depends on hope [31]. She adds, "Trust links reliance with responsibility." We need mutual trust. On the crucial role of hope, she says: "morally reparative measures must often aim at restoring or igniting hope." See also Wilburn's review [32].

A relationship also approaches remission when there is a growing emotional distance between two parties, when repeated conflicts escalate, when there is a stalemate, or when there is an addiction or mental health problem and the party concerned is reluctant to address it.

Two people may become good friends because of a similarity in their worldviews. If the these views deviate too much, then the friendship may not be worth repairing. It might happen that later their dispositions become closer again, and the friendship may or may not be repaired.

It's worth thinking about letting go when the foundations are rotten. Changes at the margins are not helpful if the core or essence of what needs to be fixed is broken. A great actor cannot make a bad movie a good movie. A bad architect's work is difficult (and/or very expensive) for a good architect to fix. If the basics are bad, repair cannot be good or effective.

The Concorde effect is associated with the sunk-cost fallacy [33]. It occurs when we are attached to something in a harmful way only because we have already invested energy, time, and money in it. The concept is named after the decisions (and expensive consequences) of the French and British governments. In the early 1960s, the French and the British jointly began to develop a supersonic aircraft that could carry passengers from Europe to America at incredible speeds. Before the first Concorde was completed, analysts predicted the project would be unprofitable [34]. It turned out that the costs would never be recovered, yet both governments poured billions of additional dollars into it. For nearly three decades, they subsidized the loss until a crash in the summer of 2000 dashed all hope.

The Transformation of Functions

As time passes, the structure of our life changes, and the objects and relationships we bring from previous eras and situations play different roles. In other words, we no longer need to repair what is damaged because our situations, needs, and wants have changed over time.

I (Z) may repair the fading color of my carefully kept, old-fashioned, hippie-style blouses by redyeing them, or the tears by sewing them. But it may be worth passing them on to a younger woman, whose life may still have an organic need for such a brightly colored or fancily cut piece. Such is the case when a parent feels his or her 20-something child growing distant and wants to mend the relationship with a Santa Claus parcel or a chocolate Easter bunny. But an adult child's relationship with his or her parents is different from that of a small child: A chocolate bunny is not a time machine.

When Objects Have Served Us

I love my old, dear, much-used hiking boots, which I have worn on many Alpine trails. We have "seen" the most beautiful mountains of Hungary together. If the worn soles make walking uncomfortable, I have to let them go, thanking them for their service. Even if the old neighbor was nice to me when I moved away, we don't have to keep in touch if the relationship did not turn into a friendship: As a neighbor, we have served each other.

If I Fix It, It Will Still Be Bad

We have old machines that consume a lot of electricity but still do not work as efficiently as modern ones. The relationship we have with these objects is toxic: They consume a lot and give nothing. This includes items that are not aesthetically repairable, such as furniture that is too ugly to look at.

To Repair: At What Cost, for What Purpose?

Most important: If it takes fewer resources to repair than to buy something new, but the difference in use is not significant, this is the easiest category to tackle. The aims of the Right to Repair movement, described earlier, are about this precisely. Sometimes you need to call in a professional to help you make the right decision.
It'll Never Be the Same as It Was, but Still …
Something that has been repaired is sometimes a joy just to look at. Maybe my grandfather's broken-handled knife, repeatedly mended, will never cut with the same vigor. It has to be handled carefully and is therefore slower to use. But when I hold it, and every time I touch the glued wooden handle, I am delighted because it reminds me of my grandfather (born at the very end of the nineteenth century) and the whole of the twentieth century. I know it was worth it to repair it.

A repaired object also reminds me of the lessons of ruin. We can generate a serious misunderstanding with a colleague in a few minutes. It takes hours to thoroughly clarify the misunderstanding, but it is worth it because our cooperation is more valuable in the future and we can draw on the resources of our relationship if necessary. Even bigger misunderstandings and conflicts can be tolerated in a jointly nurtured working relationship.

An old book of mine had fallen apart into many pages, and it was only with significant patience that I was able to glue it together so that the end result could be called a book again. But I did it, and it was good to ponder the idea that the whole is greater than the sum of its parts. If I had put the pages in a wooden box, even in numbered order, it still would not have been a book.

A change in an object's use can create new value: There are broken cups that can be glued back together but never in such a way that they can be used for tea. They still have a function, however, when placed on the dresser, and they remind me of the tea service of old. If we put the broken pots in the same place, they would represent decay, the past, and thus preservation, remembrance. There are multiple possibilities for "repurposing," and we should select the one most appropriate for our needs. It does not make sense, say, to break the pot into small pieces, glue the pieces into a single block, and declare it a paperweight.

If something is not ruined, just worn, it is still worth repairing. A nice leather bag, or meeting your former teacher every five years at reunions, for example, might both constitute examples of wear and tear that can be mended. Vintage items have the potential to become antique, and no one questions the value of antiques anymore.

5.2 Transition to the New Normal: Some Case Studies

There are different pathways for treating unstable or undesirable situations. In the previous chapter, we discussed some pathways that lead back to what we considered normal. In this chapter, we have studied some mechanisms that help the transition toward a "new normal."

Ancient Book Repairing in China

Zsuzsa started working at The British Council Library in Budapest shortly after graduating in the early 1990s. She loved the sunny building in the city center, the huge windows, the transparent, bright library space, and its interested readers and tireless, friendly colleagues. She also loved when she could retire to the storeroom to repair an expensive and important book. She did not do this work as a specialist conservator (although she was a librarian by then). She used only simple adhesive and scissors. The aim was to make the volume last as long as possible, so that she would not have to order new copies as often—like those of Francis Bacon, which are always very popular.

At the same time, Zsuzsa tried to make sure that the gluing was neat and that the book was easy to turn. She enjoyed the meticulous work and the way the books were essentially renewed in her hands. For this reason, she was delighted to read decades later, in the autumn of 2021, that the repair of ancient Chinese books is now entrusted to so-called "book doctors." One of them, the increasingly well-known Lian Chengchun, who learned the profession as an apprentice to book repair master Du Weisheng at the National Library of China, now has her own antique book restoration studio in Beijing and even teaches the trade.

Lian is a true artist. She is a meticulous, masterful craftswoman: The delicate work of her fingers brings damaged, tired pieces of ancient heritage to life. She herself prepares the glue, dismantles the book, and selects the right paper and tools for repairs. She identifies the damage to each book and decides what is needed in each of the many steps and procedures of the restoration work. In the meantime, as Lian herself puts it, she feels as if she is building a relationship with the people and characters of a bygone age [35].

The work of Lian and other book doctors is invaluable today. In fact, it is estimated that there are currently around 50 million antique books in China—including books written or printed before 1912 in the classical bookbinding style. Of these 50 million, the top 20 million are protected: So there will be plenty for Lian and her colleagues to do. And given the sheer volume, it is no wonder that the first museum dedicated to the repair of ancient books opened in 2019 in southwest China's Sichuan province. The book doctor's work is therefore both a vital and an endangered craft: The knowledge of skilled, trained restorers is much needed by the next generation of professionals [36].

Kintsugi: the Golden Repair

We will review here the Japanese approach that might be called the golden repair of failures. Kintsugi is a centuries-old Japanese method of repair. The underlying idea is that the present state of an object might reflect its history. The technique of using gold to mend a broken tea bowl does not deny, but emphasizes, the beauty of breaks and imperfections. What one might see as an imperfection, Kintsugi interprets as cultural gain.

Legend has it that a fifteenth-century Japanese ruler, Shogun Ashikaga Yoshimasa, was very fond of a pale green Chinese porcelain cup. Unfortunately, the cup broke, but the shogun didn't want to throw it away: He sent it to China to be repaired. The Chinese ceramists used metal clasps to fix the broken pieces, which was not a pretty sight, and it is no wonder that the shogun was not happy. He loved his cup so much that he approached a Japanese ceramicist and asked for his help. The craftsman was struck by the shogun's persistence and his deep love for the cup and decided to create a work of art from the broken remains. He carefully glued the individual pieces together with vegetable resin, then painted the glue with gold dust after it had dried [37].

Kintsugi portrays repair as a process in which loss can be highlighted to celebrate synthesis and rebirth. It is not just a procedure, methodology, technology, or process, but also an approach and an attitude.

The art of Kintsugi not only rehabilitates the process of injury and repair, but also makes it unrepeatable, personal, and unique. This philosophy can apply not only to a broken teacup, but also to a relationship gone awry, a reckless deforestation, or a poorly designed curriculum.

The art historian Kelly Richman-Abdou writes about it in this way: "In addition to serving as an aesthetic principle, Kintsugi has long represented prevalent philosophical ideas. The practice is related to the Japanese philosophy of wabi-sabi, which calls for seeing beauty in the flawed or imperfect. The repair method was also born from the Japanese feeling of mottainai, which expresses regret when something is wasted, as well as mushin, the acceptance of change" [38]. Once again, it is not just repairing. Rather, it is creating something more valuable than the original—from broken ceramics and objects.

As Eastern philosophies have become more widely known and accepted in the Western world over time, so has the popularity of Kintsugi. It is about the beauty in imperfection and the praise of imperfection and simplicity. Kintsugi teaches us to appreciate the signs of wear and tear that come from using objects and accept the flawed or imperfect. It helps us experience life through the senses and not to dwell on unnecessary thoughts—true understanding is not achieved through words and language anyway.

The justification of Kintsugi in our modern world is less and less questionable. Pu-Ying Lorelei Kwan writes about emotionally durable design, stating, "It is time for designers to take a step out of industrial, economic based design to one that can create or accelerate a change in the behaviour of customers …When users are not involved in the design and the making process, it will be hard for them to be emotionally attached to the object" [39].

Through the repair of a single object, Kintsugi also teaches us that trauma is part of life, and that if we work through it, we become stronger and more unique through our injuries and losses. This approach can be linked to the Japanese philosophy of mushin (no mind), which is characterised by non-attachment and acceptance of change.

Today, the Kintsugi philosophy has an even broader relevance: It can be a source of comfort for those struggling with loss, illness, trauma, and disruption to daily life during the COVID-19 pandemic. In a 2020 article entitled "Life After COVID-19: Making space for growth," Kirsten Weir wrote: "In the traditional Japanese art of kintsugi, artisans fill the cracks in broken pottery with gold or silver, transforming damaged pieces into something more beautiful than they were when new. Post-traumatic growth is like kintsugi for the mind" [40].

Esfahani Smith, a writer and journalist based in Washington, D.C., wrote about Kintsugi as a symbol or metaphor for rebuilding after traumatic events: "Kintsugi was later embraced outside Japan as a philosophy for living: Bad things can happen that might shatter us. But we don't have to stay broken or hide our wounds. We can put ourselves back together, and the scars we wear at the broken places become a reminder of the tragedies we've endured and how we overcame them—a mark of beauty in an imperfect life" [41].

So Kintsugi's approach is not just about making things better. The object is reborn, and we possess it again, even if in a changed quality. Our relationship to the object is reborn, and in this relationship we ourselves are reborn. We experience the process of restoration and accept that things will never be the same as they were. But they can still have a more beautiful, even fulfilling, role in our lives.

Worries and Hopes: A Balanced View

The Pew Research Center ran a survey among 915 innovators, developers, business and policy leaders, researchers, and activists about expectations for the "new normal" of 2025 [42]. The sample is certainly not representative, but the results tell us something important: The responses reflected a more or less *balanced* view of whether life will be better or worse. On the "worries" side of the spectrum were views related to worsened economic inequality, greater power wielded by big technology firms, and the spread of misinformation that will manipulate public perceptions, emotions, and actions. However, there seem to be hopes too. Respondents noted optimism about new reforms aimed at racial justice and social equity, enhanced quality of life for many families due to new workplace arrangements, and technological progress that contributes to better lives and safer environments.

There is only one Earth and one human race. It is very important to cooperate on global problems, even as competition remains the driving force of success.

5.3 Lessons Learned and Looking Forward

When people use the term "new normal," they express *two* messages. Trivially, first, the old is over. More importantly, second, you do not have to worry, at least there will be *something*. So, it is not the end of the world, but things will be different. You should adapt to the new situation.

Theories about sudden changes in living and social systems emerged from those about changes in the hard sciences. Self-organization is a mechanism for maintaining and forming patterns in time and space. Both mathematicians (like Alan Turing) and social scientists (like Thomas Schelling) knew that local interactions lead to global ordered patterns, even without getting instruction "from the boss."

Many natural and socio-technical systems operate close to critical points, places where small changes can have dramatic consequences. Please note, the term "change" is neutral, it may spell disaster or redemption. While we gave a number of illustrations of changes to demonstrate the power of the transition to new normal, our main message is to show the possible mechanisms of the transition between the Throw-away Society and the Repair Society.

In the next chapter we discuss the concept "Repair the World!" The concept is being studied from an historical perspective, but we use it to propose the framework of an action plan for human society.

References

1. Thom R (1972) Structural stability and morphogenesis. CRC Press, Boca Raton
2. Érdi P (2007) Complexity explained. Springer, Berlin
3. Murray SR (2022) The rise and fall of catastrophe theory. Encyclopedia.com. https://www.encyclopedia.com/science/encyclopedias-almanacs-transcripts-and-maps/rise-and-fall-catastrophe-theory. Accessed 17 Apr 2022
4. Zeeman EC (1977) Catastrophe theory: selected papers, 1972–1977. Addison-Wesley, Boston
5. Zahler RS, Sussman HJ (1977) Claims and accomplishments of applied catastrophe theory. Nature 269(10):759–763. https://doi.org/10.1038/269759a0
6. Arnold VI (1992) Catastrophe theory, 3rd edn. Springer, Berlin
7. Haken H (2008) Self-organization. Scholarpedia 3(8):1401. http://www.scholarpedia.org/article/Self-organization
8. Strogatz SH (1994) Nonlinear dynamics and chaos: with application to physics, biology, chemistry and engineering. Perseus Books, New York
9. Jones A, Strigul N (2021) Is spread of COVID-19 a chaotic epidemic? Chaos Solitons Fractals 142:110376. https://doi.org/10.1016/j.chaos.2020.110376
10. Ouyang Q, Swinney HL (1991) Transition from a uniform state to hexagonal and striped Turing patterns. Nature 352:610–612. https://doi.org/10.1038/352610a0
11. Murray JD (1988) How the Leopard gets its spots. Scientific American, pp 80–87. https://www.scientificamerican.com/article/how-the-leopard-gets-its-spots/
12. Schelling T (1971) Dynamic models of segregation. J Math Soc 1(2):143–186. https://doi.org/10.1080/0022250X.1971.9989794
13. Vieira T (2017, Nov 29) Inequality ... in a photograph. The Guardian. https://www.theguardian.com/cities/2017/nov/29/sao-paulo-injustice-tuca-vieira-inequality-photograph-paraisopolis
14. Hoque L (2018, May 24) Revamping favelas: top 10 facts about poverty in Sao Paulo. The Borgen Project. https://borgenproject.org/revamping-favelas-top-10-facts-about-poverty-in-sao-paulo/
15. Veneri P (2018, May 24) Divided cities: understanding intra-urban inequalities. Presented at Inequality matters: Champion Mayors Webinar series on understanding & overcoming segregation in cities. OECD Champion Mayors Initiative. https://www.lincolninst.edu/sites/default/files/sources/events/inequality_matters_webinar_24_may_2018_spatial_segregation.pdf

16. Helbing D, Johansson A (2009) Pedestrian, crowd and evacuation dynamics. In: Meyers RA (ed) Encyclopedia of complexity and systems science. Springer, Berlin. https://doi.org/10.1007/978-0-387-30440-3_382

17. Helbing D, Farkas I, Vicsek T (2000) Simulating dynamical features of escape panic. Nature 407:487–490. https://doi.org/10.1038/35035023

18. Hegselmann R, Krause U (2002) Opinion dynamics and bounded confidence: models, analysis and simulation. J Artif Soc Soc Simul 5(3). https://www.jasss.org/5/3/2.html

19. Carothers T, O'Donohue A (2019) Democracies divided: the global challenge of political polarization. Brookings Institution Press, Washington, DC

20. Wright M (2021, Sept 2) Kaizen: the Japanese approach to continuous improvement. KaiNexus. https://blog.kainexus.com/improvement-discipline/kaizen/kaizen-the-japanese-approach-to-continuous-improvement

21. Stoll J (2021, Nov 4) Share of adults with a Netflix subscription in the United States as of August 2021, by generation. Statista. https://www.statista.com/statistics/720723/netflix-members-usa-by-age-group/

22. Catloth J (2019, Dec 7) Creative destruction and Netflix. Bearmarket. https://bearmarketreview.wordpress.com/2019/12/07/creative-destruction-and-netflix/

23. Gladwell M (2006) The tipping point: how little things can make a big difference. Little, Brown and Company, Boston

24. Thresholds and tipping points. University of Southampton. http://www.complexity.soton.ac.uk/theory/_Thresholds_and_Tipping_Points.php. Accessed 17 Apr 2022

25. Kuhn T (1962) The structure of scientific revolutions. University of Chicago Press, Chicago

26. Dosi G (1982) Technological paradigms and technological trajectories: a suggested interpretation of the determinants and directions of technical change. Res Policy 11(3):147–162. https://doi.org/10.1016/0048-7333(82)90016-6

27. Ma A (2022) Repairing society. Andy Ma Design. https://www.andymadesign.com/repairingsociety. Accessed 17 Apr 2022

28. Eldredge N, Gould SJ (1972) Punctuated equilibria: an alternative to phyletic gradualism. In: Schopf TJM (ed) Models in paleobiology. Freeman Cooper, pp 82–115

29. Gersick CJG (1991) Revolutionary change theories: a multilevel exploration of the punctuated equilibrium paradigm. Acad Manag Rev 16(1):10–36. https://www.jstor.org/stable/258605

30. Gersick CJG (2019) Reflections on revolutionary change. J Change Manag 20(1):7–23. https://doi.org/10.1080/14697017.2019.1586362

31. Walker M (2006) Moral repair: reconstructing moral relations after wrongdoing. Cambridge University Press, Cambridge

32. Wilburn B (2007, May 9) Moral repair: reconstructing moral relations after wrongdoing. Notre Dame Philos Rev. https://ndpr.nd.edu/reviews/moral-repair-reconstructing-moral-relations-after-wrongdoing/

33. Ronayne D, Sgroi D, Tuckwell A (2021, July 15) How susceptible are you to the sunk cost fallacy? Harvard Bus Rev. https://hbr.org/2021/07/how-susceptible-are-you-to-the-sunk-cost-fallacy

34. Arkes HR, Ayton P (1999) The sunk cost and Concorde effects: are humans less rational than lower animals? Psychol Bull 125(5):591–600. https://doi.org/10.1037/0033-2909.125.5.591

35. Hongyu LJ (ed) (2021, Sept 13) Chinese woman makes ancient books shine again with her exquisite handiwork. People's Daily Online. http://en.people.cn/n3/2021/0913/c90000-9895479.html

36. Xinhua (2019, Dec 27) China opens first museum on ancient book repairing. ChinaDaily.com.cn. http://global.chinadaily.com.cn/a/201912/27/WS5e056b5ca310cf3e35581042.html

37. Kintsugi (2022) UnmissableJAPAN.com. http://www.unmissablejapan.com/etcetera/kintsugi. Accessed 17 Apr 2022

38. Richman-Abdou K (2019, Sept 5) The centuries-old art of repairing broken pottery with gold. My Modern Met. https://mymodernmet.com/kintsugi-kintsukuroi/

39. Kwan PYL (2012) Exploring Japanese art and aesthetic as inspiration emotionally durable design. Presented at DesignEDAsia conference 2012. https://www.semanticscholar.org/paper/EXPLORING-JAPANESE-ART-AND-AESTHETIC-AS-INSPIRATION-Kwan/3836ebc85bd632d20f36d747f796075d1a2c2ccf
40. Weir K (2020, June 1) Life after COVID-19: making space for growth. Monit Psychol 51(4). https://www.apa.org/monitor/2020/06/covid-life-after
41. Smith EE (2021, June 24) We want to travel and party. Hold that thought. The New York Times. https://www.nytimes.com/2021/06/24/opinion/covid-pandemic-grief.html
42. Anderson J, Raine L, Vogels AE (2021, Feb 18) Experts say the 'New Normal' in 2025 will be far more tech-driven, presenting more big challenges. Pew Research Center. https://www.pewresearch.org/internet/2021/02/18/experts-say-the-new-normal-in-2025-will-be-far-more-tech-driven-presenting-more-big-challenges/

Chapter 6
Repair the World!

Abstract Since ancient times, the concept of "Repair the world" has been with us. At the individual level, people have had the wish for rebirth or at least renewal. At the global level, we need self-stabilizing technologies and multilateral negotiations to avoid the threat of existential risks due to technological catastrophes or international conflicts. We have the catastrophic legacy of global injustice, and while history can not be reversed. The new generation should learn about the past from different perspectives. It is possible to avoid the "tragedy of commons", but we need peace.

6.1 From Ancient to Modern Perspectives

6.1.1 Mythical Origins

Tikkun Olam
The concept of "repair the world" has evolved from ancient to modern times. Historically, according to the early rabbinical literature, *tikkun olam* (repair the world) implies that while the world is innately good, its Creator purposely left room for us to improve upon His work. As the Canadian Rabbi Tzvi Freeman cites:

> If It's Broken
> If you see something that is broken, fix it.
> If you cannot fix all of it, fix some of it.
> But don't say there is nothing you can do. Because, if
> that were true, why would this broken thing have come into your world?
> Did the Creator then create something for no reason? [1]

According to a school of Kabbalah (the ancient Jewish tradition of mystical interpretation of the Bible), Adam, the first man, was intended to restore the divine spark. However, good and evil remained mixed in the world as a result of his sin, and human souls (previously contained within Adam's) also became imprisoned. The prophets frequently talk about the need to create an ethical society. Even their wording sounds like a contemporary *tikkun olam* manifesto. As is known, the Torah speaks endlessly about loving the stranger and the poor. The present-day usage of the phrase emphasizes that the world is profoundly broken and can be fixed only by human activity. It is interpreted as the Jewish approach to social justice in the secular world. The term *tikkun olam* is connected with human responsibility for fixing what is wrong with the world [2].

It is not so important whether today's *tikkun olam* reflects Judaism. What counts is that people think about the omnipresence of the concept of repair. If our first thought is "Can I save it?" then we have made a step toward living in the Repair Society. In any case, the concept of "repair the world" is as old as recognizing that the world is not perfect. Repair is related to renewal and rebirth, at least in the mythical sense.

Rebirth in Mythology

"It's like I've been reborn!"—one sometimes cries out after medical treatment, a holiday, an exciting event, or even after entering into a new relationship. When we are concerned about a particular subject or question, it is always worth seeing if and how Greek mythology relates to it. This vast collection of stories from human culture sheds light on psychological and philosophical connections that are still valid today. In mythology, the motif of renewal, resurrection, and rebirth often appears in close coexistence with nature. Below are some examples of mythological figures [3].

One of the most influential figures in the concept of renewal and rebirth is Asclepius, son of Apollo and Coronis, god of healing and medicine, whose daughter Hygieia is the goddess of health. Asclepius learned the science of recovery from the centaur Cheiron on Mount Peleon and perfected it to such an extent that he could cure all diseases, heal all wounds, and raise the dead. He owed this special ability to Pallas Athena, the goddess of wisdom, who gave him two vials of blood. (The vials contained the blood of Medusa, whom Pallas Athena had previously angered). The sacred animal of Asclepius, the serpent, is the symbol of renewal. The way the snake bites into its tail symbolizes constant renewal and eternal cycles. The myths reflect not only the renewal of man but also of nature.

The phoenix, known to many from the *Harry Potter* series or the *Chronicles of Narnia* films, is a creature of Egyptian mythology. It is sometimes depicted as a stork, sometimes as an eagle or peacock. According to some accounts, the beautiful, golden-red feathered bird may have lived for more than 10,000 years: It burns to ashes, and a new phoenix is born from the ashes. It is a symbol of rebirth, as well as of the sun god or heavenly grace, resurrection, and eternal life. It was believed that if a woman dreamed of a phoenix, she would give birth to an excellent son.

It is no coincidence that mythology depicts the idea of rebirth in such detail. This process—or moment—gives humans the opportunity, again and again, to transform, repair, and renew something in our lives. In this sense, it is also the promise of change and hope, which the world offers to us and we offer to ourselves.

While the myths of renewal may give some cosmic hope that destruction is not final, we should use our rational minds to actively avoid the world's destruction.

6.1.2 How to Avoid Existential Risk?

Nuclear War

Nuclear war between the United States and the Soviet Union was perceived to be a real threat during the Cold War. The Cuban Missile Crisis of 1962 is generally considered the time when the world was the closest to an outbreak of nuclear war. There was a general belief that both countries could destroy the other's infrastructure and a large proportion of its population.

As the best seller book and movie *Fail Safe* mentioned in Sect. 3.2.3 suggested in the early 1960s, a nuclear war might emerge due to technological glitches [4, 5]. In the following decades, it happened many times that the malfunction of early warning systems led to false alarms: "Not once, or twice, but frequently. There is no publicly available real history of all the failures. However, it is known that between 1977 and 1984 the US early warning system showed over 20,000 false alarms of a missile attack on the US. Over 1000 of these were considered serious enough for bombers and missiles to be placed on alert" [6].

At the same time, bilateral (and later multilateral) agreements reduced the danger of nuclear war. Here is a reason to have cautious optimism: "Five of the world's most powerful nations have agreed that 'a nuclear war cannot be won and must never be fought' in a rare joint pledge to reduce the risk of such a conflict ever starting" [7]. However, nuclear terrorism by non-state organizations remains a source of unpredictable danger.

Technological Singularity

The concept of technological singularity describes a point in time at which humans can no longer control technological development. It is often associated with the notion of *exponential technological growth*. However, the terminology is somewhat misleading for two reasons. First, as we learned earlier from Fig. 3.1, the condition for having finite-time singularity is super-exponential growth. Second, unbounded exponential growth does not often happen. Generally, there is are negative feedback mechanisms that "pull back" growth.

Moore's Law was an observation-based hypothesis developed by Gordon Moore, co-founder of chip manufacturing giant Intel. He predicted in 1965 that the number of transistors on a microchip would double every year (later revised to 18 months). More importantly, the power and speed of computers would also increase exponentially, with a doubling time of 18 months. Many other technologies—such as 3D printing, drones, robotics, artificial intelligence, and synthetic biology—show higher-than-linear growth. Now we have smartphones, tablets, readily available Wi-Fi, GPS, social media, and many other technologies unimaginable just a few decades ago.

Daniel Schmachtenberger, a social philosopher, has sounded the alarm about the existential risks associated with exponential technological growth [8]. Some of his theses can be interpreted as follows:

- There is a contradiction between local and global. Things that work well for individuals locally are directly against global well-being. Powerful technologies, which initially look useful, later may be destructive. In a game-theoretic sense, we play a win-lose game with the chance that everybody will lose.

- Sustainable systems are closed-loop and self-stabilizing and can repair themselves. Man-made systems, like industry, are historically open-loop systems. They create waste and must lead to collapse.
- The goal is to create loop closure in man-made systems based on circular technologies.
- Society needs to transition from its current rivalrous game (in his terminology Game A) to a sustainable, win-win game (Game B). We should find out how to convert the Game A we play to Game B [9].

Well, the expression *needs* is normative. Moral philosophy could and probably should prescribe norms. However, science should help to find realizable mechanisms to reach normative goals.

How to Repair the Legacy of Past Global Injustice

We live in a world inherited from our ancestors. We don't want to state that the newer generations are innocent, but different versions of structural discrimination (such as racism and white supremacy) were initiated centuries ago. We know how European colonialism led to the collapse and the loss of Aztec, Inca, and other cultures. It resulted in the loss of over 80% of indigenous populations. The killing of George Floyd in May 2020 dramatically accelerated public awareness of persistent systemic racism. What is lost is lost; processes are irreversible; we cannot reverse history. But some damage control is possible. The harm caused could and should be repaired, and we can study the mechanisms which led to these injustices in order to avoid repeating them.

One of the big battle fields of social repair movements is that of post-colonialism. How many experiences of colonization and how many reparations are we contending with? Dealing with the imperial past raises difficult and divisive questions. King Philippe II of Belgium recently expressed regret for his ancestor's brutal regime, which caused the deaths of an estimated 10 million people. Later that day, the bust of a former monarch Leopold, responsible for the death of millions of Africans, was removed. We leave it for the Reader to decide whether these events can be seen as repairing historical damage. It is difficult to compare the damage caused in different countries and regions—and quantifying the cultural or emotional losses is almost impossible. Multidisciplinary expert work—and sensitive diplomacy—is needed to ensure that the victims and their descendants (who may equally be victims of indirect losses) receive fair reparations for the damage caused by the various evils of colonialism.

One of the most straightforward steps seems to be to return the stolen treasures. Or is it not so straightforward? In the summer of 2016, the Beninese government, with the support of the Conseil Reprèsentatif des Associations Noirs, asked France to return treasures stolen during the November 1892 conquest. A few months later, in December, they received a negative reply. Finally, under pressure from African countries, France acknowledged the injustice and in December 2020 passed a law to return stolen cultural property to Benin and Senegal. Similarly, in the spring of 2017, the Museum of Anthropology and Archaeology at the University of Cambridge announced an agreement to loan (so as not to have to return) to Nigeria the treasures stolen during the conquest of the Kingdom of Benin (now Nigeria) at the end of the nineteenth century. These steps are not so simple. Often, it has been so long since the treasures were stolen that returning them can seem a painful farewell for the former colonizers. The process of reparations and apology is effective when neither party feels upset by the delicate balance of the situation [10].

The debate is not just in France and England, and it is not just about treasure. Reparations also involve acknowledging the damage caused by colonial wars and violence and healing the wounds of historical cruelty. It is not easy for today's European nations to acknowledge the destructive role they played in former colonies and the long-lasting consequences of that role. Moreover, recognition is far from justice. It is a question for the future how and by what steps the process of reparation can be continued [11].

There are still obstacles, despite precedent for victims and descendants of historical abuses demanding reparations. One of the biggest questions is how reparations—their nature, their extent, their timing—can be determined in a way that helps society heal and does not create further divisions. Further divisions can result if, for example, the value of reparations is "insultingly low." Or if reparations exclude groups that may have suffered other types of harm. Estimating damages is also a difficult task, and subjectivity is bound to emerge, which may prevent agreement on acceptable settlements. Once the damage has been estimated, the next question arises: How much money can be spent to compensate for the loss? How can the money be distributed among different aggrieved groups? In other words: Who faced greater challenges, who suffered more, who should receive more money?

Adom Getachew, a professor of political science at the University of Chicago, writes about the new political reality of decolonization. Between 1945 and 1975, when independence struggles were fought in Africa and Asia, decolonization was primarily political and economic, and United Nations membership grew from 51 to 144 countries. Then decolonization began to gain momentum, in parallel with a recognition of Eurocentrism that identified indigenous cultural traditions and knowledge systems as backward and the colonized as people without history. Students at the University of Cape Town were already calling for an Africa-centered curriculum in the second half of the 2010s. Getachew points out that recent decolonization movements have emphasized that colonialism not only shaped the Global South but also created Europe as we know it today and the modern world. The profits from the slave trade, for example, supported the rise of port cities such as London, Liverpool, and Bristol, while building a narrative of African underdevelopment and marginalizing the role of violent exploitation as much as possible. The protest movements of recent years have shattered not only statues but also illusions. The illusions are replaced not by a redrawn past but by a present realized in reparations. Who owes what to whom is the subject of lively debate around the world. But we cannot forget that reparations on this scale are not one-off transactions, but part of a process of transformation [12].

Maybe New Zealand provides some hope. In addition to material treasures, ideas and culture were lost to colonialism. The New Zealand government has recently proposed that schoolchildren should learn about the history of the indigenous Maori people and British colonialism in a specific curriculum, consolidating the knowledge of colonialism and reflecting a renewed appreciation of Maori culture and history. Critics of the draft curriculum say it lacks balance, for example by emphasizing the country's short history of achieving one of the highest living standards in the world. True, this does not change the fact that the Maori, who make up around 15% of New Zealand's population, were dispossessed of much of their land during colonialism. It is not easy to create and implement measures that adequately support improvements in living conditions without grappling with these fundamental inequalities and their relationship to colonialism [13].

The *catastrophic legacy of past global injustice* suggests that untreated social tensions increase the vulnerability of today's society. We now continue our analysis to connect past and future. In speaking about restoration, let's step back from the social realm to that of objects.

6.2   From Times Past to Modern Theory to Action

The Ethics of Restoration

From antiquity to the present day, many considerations have guided those who touch objects [14]. Three main considerations stand out:

1. We may want to use objects again, in their original function, and therefore "improve" them. The act of repair is as old as the toolmaker himself. Good examples are Bronze Age pots, on which patches were fixed by riveting, or ceramics, fragments of which had holes drilled in them and were then fastened together with iron or copper wire.
2. We may want to restore the aesthetic beauty of an object, but we may do so according to modern aesthetic criteria and not with the goal of restoring an object's original condition.
3. We may, as in the practice of conservation-restoration, aim to restore the aesthetic and historical integrity of a damaged, deteriorated, or aging artifact and to apply conservation techniques that will ensure its survival in the future.

In Europe, object conservation probably became a conscious activity during the Renaissance, when the collection and display of antiquities became commonplace. The restorer has the extreme responsibility of working on irreplaceable original works of art of great artistic, historical, ethnographic, scientific, cultural, social, or commercial value. Whether working alone or as part of a team, the conservator's responsibility is to balance the conservation of the artifact and society's need to preserve its cultural assets. They must seek to understand the power of art and its cultural value.

Under the Spell of Fragments

When, in the 1820s, Lord Elgin sold friezes from the Parthenon in Athens to the British Museum, they were not restored. They were shown to the public in the fragmentary form in which they were found. Previously, when fragments were found, almost everything was restored. After the decision regarding the friezes, fragments became a cult-like phenomenon that still appear in Rilke's poem "Archaic Apollo's Torso," 100 years later:

> We cannot know his legendary head
> with eyes like ripening fruit. And yet his torso
> is still suffused with brilliance from inside,
> like a lamp, in which his gaze, now turned to low,
> gleams in all its power …[15]

As we can see in the verse, the fragment is peculiar in that the missing elements make the overall effect even more dramatic.

The aesthetics of incompleteness are felt in this poem. In Rilke's eyes, this torso carries and radiates its entire self through the power of art. Rilke makes an absent, invisible, and eternally lost gaze visible. He brings existence to nothingness in his poem. It is particularly brilliant because, as the poem itself shows, the statue is missing its head, but Rilke sees a radiant gaze come to life.

The cult of the fragment was first truly peculiar to Romanticism, which idealized naturalness, individuality, and, consequently, often imperfection, in contrast to the controlled Baroque and Classicism. The period was also characterized, for example, by the art-roman as an architectural form, where elements such as the English garden were created to directly resemble buildings that had been weathered by time. Here, the fragmentary is a way of looking to the past, of idealizing the past, because what is missing adds to the overall picture. They didn't want to overwrite the past but to leave it as time had eroded it.

In the nineteenth century, therefore, the art of the past and the present were separated. As a result, the art of the past becomes a great work to which you can neither add nor take away.

If we don't have an entire object, maybe we can't do anything better than to appreciate the fragments. But it would be an exaggeration to state that fragility itself is a desirable feature of a system. Just the opposite! The pandemic has demonstrated the fragility of our world. So we may want to see and design anti-fragile systems.

Taking (Anti-)Fragility Seriously

Heinz von Foerster (1911–2002), born and raised in Vienna, served as the secretary of the last five Macy conferences, the celebrated meetings on cybernetics. Between 1958 and 1975, he also directed the very influential Biological Computer Laboratory at the University of Illinois at Urbana-Champaign. He famously constructed and defended the concept of second-order cybernetics, which emphasized autonomy, self-organization, cognition, and the role of the observer in modeling a system. Cybernetic systems, such as organisms and social systems, are studied by another cybernetic system—namely, by the observer. von Foerster gave a simple physical example (interaction among magnets) as an illustration of the principle he called *order from noise* [16]. From this, we learned that noise or random perturbations could help a self-organizing system find new normal stable states.

A half-century later, Nassim Taleb formulated a more general idea and suggested that noise, in addition to stressors, shocks, volatility, mistakes, faults, attacks, or failures, might have positive effects in specific systems, which he labels anti-fragile [17]. He does not cite von Foerster, but the different traditions are in conversation with one another.

For anti-fragile systems, uncertainty and risk are positive features. The classic example of anti-fragility is the Hydra, the Greek mythological creature with numerous heads. When one is cut off, two grow back in its place. If you have a very stable system, say technological systems that proved to be very reliable over the years, an unexpected failure might be especially bad because nobody has hands-on knowledge of how to repair it. But if it breaks a lot, you will learn how to fix it. According to an optimistic perspective, "if it breaks, even more, you may create systems that know how to fix it: An auto-repair system" [18]. In the "socialism" in which we grew up, devices often became disabled. So both of us saw many craftsmen who developed incredible skills to repair almost everything, even in a shortage economy. But instead of designing an auto-repair system, the whole political economic system fell to pieces.

Stress is an essential element of anti-fragility. Just as focused exertion can help build muscles and bones, environmental stress may (or may not) help organizations evolve and adapt to potential challenges. Taleb believes that *modern* society is fragile since it wants to control and predict everything.

How should we characterize the anti-fragile society? It is not a new idea, but it is still good to remember that closed societies built on top-down, hierarchical systems of control and rule-based order try to stabilize the status quo. Generally, an ideological dogma or belief system provides the operational framework. The goal is to act based on planning. Open societies accept the co-occurrence of both order and uncertainty. They try to unify constancy and change. (Somewhat analogously to the first and second law of thermodynamics, which we discussed in Sect. 3.1.) New ideas naturally create uncertainties. It is good to accept that there is no perfect order and no absolute truth, not even as an asymptotic goal. In a closed society, a new problem is seen as an adverse event. In an open society, it is seen as a possibility for renewal. Somewhat paradoxically, open societies are therefore more stable. Autocratic closed systems show more instability, with more riots and subversive activities [19].

Since we believe that any open society should be based on technological progress and good morals, we now discuss the most critical perspectives regarding technological changes.

Action: Toward Restorative Justice

Here we turn to lessons learned from the modern criminal justice system and discuss briefly the age-old question of balancing punishment and repair. Restorative justice emphasizes repairing and healing people who have been harmed. Evolutionary psychologists argue that both retribution and forgiveness are universal human adaptations that have evolved as alternative responses to exploitation and proved to be useful strategies for reducing the risk of re-offending [20].

Restorative justice is a different way of thinking about crime and our response to crime, and it focuses on attempting to *repair* the harm caused by crime and reducing future harm through crime prevention. According to this approach, offenders should take responsibility for both their actions and the harm they have caused (again, a normative rule). Restorative justice seeks compensation for victims from offenders and the reintegration of both within the community. It also requires a cooperative effort by communities and the government.

The functional principles of restorative justice have been summarized as follows [21]:

- Crime causes harm and justice should focus on repairing that harm.
- The people most affected by the crime should be able to participate in its resolution.
- The responsibility of the government is to maintain order and of the community to build peace.

The main building boxes of restorative justice are:

- Inclusion of all parties
- Encountering the other side
- Making amends for the harm
- Reintegration of the parties into their communities.

In the traditional justice system, victims of crimes rarely have the opportunity to communicate with the offenders who harmed them. Restorative justice is based on the assumption that communication will not create more harm. However, in the case of violent crime, there is a fair chance that direct communication will open old wounds and activate raw emotions, as is discussed in a paper about the pros and cons of restorative justice [22].

Action: Transition to a Circular Economy

The circular economy and circular technologies are not silver bullets, but they can change our thinking about waste materials. To make a large-scale transition from linear to circular technologies, we need *new technologies* and *good morals*.

The traditional linear $extract \rightarrow produce \rightarrow use \rightarrow dump$ model of modern economic systems is unsustainable. The circular economy instead emphasizes the reuse, remanufacturing, refurbishment, repair, cascading, and upgrading of components, materials, and even products. It also considers the use of solar, wind, biomass, and waste-derived energy. Ecological economics recognizes that the environment has both local and global limits. Ecological economists try to understand global issues such as carbon emissions, deforestation, over-fishing, and species extinction. They also understand the conflict between short-term policy and long-term visions of sustainable societies [23, 24].

Linear technologies convert crude materials to products and waste substances. In an ideal circular world, waste substances should be converted to crude materials through closed-loop technologies. We turn now to review some examples of circular solutions [25]. The significant step in each of these technologies is when the loop is closed.

- *Clothing: from textile to fibers* As we discussed in Sect. 1.2.4, the very linear model of the fast fashion industry persuades consumers to buy the latest styles. People living in the throw-away society don't see any problem with disposing of clothing purchased a year ago. Used clothes sent to landfills or burnt are certainly a significant component the climate catastrophe. According to the Secondary Materials and Recycled Textiles Association, 95% of textiles can be recycled, but 85% end up in landfills. What are our options for avoiding this scenario? While buying and selling clothes at the second-hand market reduces waste, it will most likely not become a mainstream business model. There are now more textile recycling programs, perhaps in your own town, that help implement better outcomes for the environment. As concerns technology, here is an encouraging example: the Spanish company *Recover*™ transforms textile waste into low-impact, high-quality recycled cotton fiber. You can learn about the closed-loop technology from the video in the notes [26].
- *From human waste to animal feed* Waste from humans and animals is still an everyday problem in many developed countries. In the absence of appropriate water treatment systems, wastewater can reach rivers and lakes and contaminate drinking water, an obvious source of serious disease. It is not always simple to install well-operating sewer systems since they often require a large amount of energy. Therefore, a question has emerged regarding whether biology could offer more efficient solutions. In 2019, Forbes reported that "Black Soldier Flies Are The New Superstars Of Sustainable Aquaculture" [27]. The single most important fact about black soldier flies may be that in, the larvae stage, they are extremely efficient at transforming waste into high-quality protein, and they can be used as protein additives in animal feed. Technology can help establish an inexpensive, clean, and sustainable food source for animals. However, it is too early to see whether the procedure can be scaled up and made eligible for approval by governmental agencies.
- *From paper waste to biodegradable plastics* Lignin is a beneficial organic polymer molecule that forms the cell walls in trees and plants. It is also a waste material from chemical pulping processes and a significant biomass component of paper production. The conversion of wood chips to a pulp for paper manufacturing generates huge lignin quantities. The good news is that lignin seems to substitute for petroleum-based plastics. It is not a useless byproduct but a starting material for a closed-loop technology. With its derivatives, lignin can be converted into recyclable products such as bottles, shopping bags, and straws. While the potential of lignin has been known for decades, actually applying this knowledge has not been easy. There are some problems in identifying its structure, and ongoing research aims to accelerate the development of lignin-based technologies [28].

The following two examples are still waste management but without closed-loop technologies:

- *From plastics to roads* Toby McCartney, the CEO of the British company MacRebur has a mission "to help solve two world problems; to help solve the waste plastic epidemic and to enhance the asphalt used to make our road surfaces around the world" [29, 30]. MacRebur introduced a technology to process waste plastics into asphalt for road construction. There is a secret element in the process. They use a "specially designed activator" in addition to the waste plastic.

- *From plastic bottles to bricks* Material scientists have created a technology to convert plastic bottles into bricks by identifying a set of potential additive materials that would improve the compressive strength of the bricks [31]. As a result, different versions of the technology have been adopted in several countries. For example, Gjenge Makers, a factory in Nairobi, Kenya, takes plastic waste and turns it into a brick that is five to seven times stronger than concrete. Nzambi Matee used her engineering skills to create the process, which involves mixing plastic waste with sand. The cleaned and dried bottles are cut into pieces. The plastic is melted in a drum, and sand from the bricks is mixed in [32]. The ecobricking movement emerged from a growing awareness of the scale of plastic pollution and has arisen in many countries, including the Philippines, India, and South Africa, to mention a few of them.

Utilizing the Sharing Economy

The sharing economy is related to the circular economy in that both models seek to reduce societal waste. The sharing economy is a means to improve our morals and forces us to accept that we should not necessarily own things. We see that sharing economy business models are rapidly spreading across the world, and we could consider them seeds for the cultural changes mentioned in Sect. 5.1.3. The basic idea is that every resource that is not utilized efficiently is wasted. For example, cars are parked 95% of the time, so the energy that went into their manufacturing is wasted [33].

The throw-away society adopts a model based on private property and a consumerist view of society. In the repair society, sharing platforms would be used extensively. On sharing platforms, the company does not make or own any goods; it just provides a peer-to-peer (P2P) platform to connect people and unused products or services. The platform regulates bargaining among parties and facilitates transactions [34].

Some examples of sharing platforms include: .

- Ridesharing or carsharing: Services like Lyft and Uber mean you do not need to own a car to get around but still don't have to rely on public transit.
- Coworking: As many people now work in isolation, in home offices, or in Starbucks shops, coworking has arisen as a way to share infrastructures and expenses. It also provides some community. My (P) son, Gábor, an individual contractor, uses the Kubik coworking space at the foot of the Margaret Bridge in Pest almost daily, and he likes it very much.

- Couchsurfing and Airbnb: New rental services provide alternatives to hotels. *Couchsurfing* was free for about fourteen years. It provided less privacy, but as compensation, if you were alone in a foreign city, you had people to talk to. *Airbnb* became a success story since it is often simply cheaper than staying at a hotel.
- Peer-to-peer lending: P2P lending is an alternative method of financing. People are able to obtain loans directly from other individuals without the participation of financial institutions. It is not clear whether or not these arrangements will be popular and whether the advantages will exceed the disadvantages.

We believe that the emergence and propagation of new cultural, moral, and technological changes leading toward the repair society may happen through gradual reforms. Next we will discuss some success stories of the reform movements.

6.3 A Capsule History of Reforms

Reform is certainly the most frequent mechanism of repairing the world. The role of reforms could be discussed in a separate book of its own, but we will analyze several significant ones. Luther initiated improvement in the Christian church by translating the Bible. Modern Japan emerged during the Meiji Restoration. Gandhi transformed a subcontinent as a leading figure in a peaceful resistance movement for Indian independence. The spinning wheel was the physical embodiment and symbol of Gandhi's constructive program and was seen as a device to help the country gain independence from the British textile industry. Jumping to the present, we will briefly mention the story of Greta Thunberg.

High-Speed Repair: The Reform

It's not just a shoe or a pair of torn trousers, a greying relationship or a crumbling castle that needs repair. National education, regional industry, state budgets, and even the state of the global environment also need repair. No one can magically transform these complex structures and processes overnight. Improvements, as well as substantial changes in the minds of peoples, can be achieved slowly, methodically, and deliberately. In contrast to revolution, reform introduces gradual change or modification—an innovation in an institution or a community—to promote progress and transform the situation without a complete break with the past. According to the internet dictionary, reform processes make changes in something (especially an institution or practice), in order to improve it [35].

Successful and Unsuccessful Reforms from Ancient Times

What can be reformed? Even one of the most important dimensions of our lives, the calculation of time, was once the subject of reform. Julius Caesar, often described by historians as thin, bald, frail, and sickly, but who was the most powerful man in the world half a century before the birth of Christ, rearranged the calendar [36]. His innovation divided time into 12 months and instituted leap years, and this reform has not changed significantly to this day. In the Julian calendar, a month was even named after him: July (and August after his adopted son Augustus) [37].

Not all reforms are successful. Another Roman ruler, Diocletian, who started out as the child of poor parents and later became a powerful and influential figure, tried to rebuild the shattered Roman Empire. He restructured the government, appointing four capitals with four vice-emperors. He fixed the highest price of food because of the famine. But above all, he persecuted Christians—a bad decision. After 20 years of rule, Diocletian retired as a private citizen and saw the futility of his struggle against the followers of Christ. His immediate successor, Constantine, banned the persecution of Christians in 313 B.C.E. [38].

Two German Reforms that Had Worldwide Impact

When the word "reform" comes to mind, we often think first of Martin Luther, an Augustinian monk from Wittenberg. He was outraged by abuses he considered unworthy of the church and denied that money could buy the divine grace of forgiveness. In his perspective, if it can be bought, it is not grace. God's grace cannot be mediated, Luther said, and he sought to let each man read the ideas of God for himself. He nailed his 95 Theses to the church door in Wittenberg, protesting against the sale of the divine grace of forgiveness. His style was raw but powerful, and his writings were bought and read all over Germany.

Despite the Pope's threat to excommunicate Luther, he burned the papal letter to that effect. (No wonder he was indeed excommunicated.) He and his followers then broke away from the church, and even high-ranking nobles and princes began to sympathize with the process that came to be known as the "Reformation." They were not at all sorry to see the power of church officials finally decline. The German emperor Charles V summoned the disobedient monk to appear before the imperial assembly at Worms in 1521. He expressed his willingness to recant his doctrines but only if they could be proved false according to the Bible. Those present were not in the mood for theological debate and urged the withdrawal of Luther's doctrines. Luther refused to do so and was therefore declared a heretic and excommunicated. This meant that anyone could kill him with impunity, and no one could give him shelter or food. (The same was true of anyone who bought or possessed Luther's books.) So the reformer was outlawed, but his patron, Frederick the Wise, "arrested" him, and Luther was taken to Wartburg Castle, in disguise, under a pseudonym. It was in this voluntary captivity that he translated the Bible into German, reforming the language, unifying and bringing Bavarian, Saxon, and other German vernaculars to the literary level [39].

The Reformation is not a stand-alone, singular event. The invention of the printing press, the development of the sciences and arts in the fifteenth century, and great geographical discoveries all contributed to the receptive environment in which Luther's movement was embedded. (In the second half of the eighteenth century, for example, the German Emperor Joseph II of Rome, son of Maria Theresa, was impatient to introduce his reforms. Looking back to the early twenty-first century, it is more or less understandable that his sometimes hasty measures were met with uproar and resistance.)

Otto von Bismarck, a strong-faced Prussian, was known as the Iron Chancellor. His reforms were not always greeted with undivided enthusiasm, but he did not fret. In the second half of the nineteenth century, he successfully united nearly 40 German kingdoms, city-states, and principalities into a single empire. It was unified as an economically and militarily powerful superpower, with social, health, and educational reforms that have had an impact to this day.

From the Shogunate to Modern Japan: The Meiji Restoration

There have also been particularly intensive reforms. When the teenage Mutsuhito seized power in a coup d'ètat in January 1868, an event known as the Meiji Restoration, Japan went from an isolated feudalistic island state to a colonial power with state-of-the-art institutions.

The Tokugawa era (1600–1868), which preceded the Meiji Restoration, established a centralized system and an authoritarian government and essentially excluded Europeans from the island nation, with the exception of the Dutch, who had a limited trading presence. Nevertheless, both political and educational standards were high, with thousands of schools affiliated with private universities, churches, and government offices. Industry and commerce flourished, and in the cities, culture ranged from kabuki theatre to woodcuts to haiku poetry. But in the early 1800s, the country's leaders became increasingly interested in the West, and American and European seafarers began to appear in Japanese ports. The shogunate became increasingly inflexible, while political and economic activity intensified in several provinces.

In 1854, the Tokugawa decided to open Japan to foreigners. A turbulent period began. Lower-ranking samurai demanded the expulsion of Westerners, and their violent actions upset the calm in the political centers. By the mid-1860s, Choshu had fallen into the hands of an anti-Tokugawa government, and by the end of 1868, Shogun Tokugawa Keiki wanted to establish a system in which power was shared among a council of leaders. His plan failed: On January 3, 1868, a coup in the name of Meiji handed power to a group of young, pragmatic leaders from the provinces, all nobles and samurai. They formulated their policies without any ideological or customary constraints, abolished class systems, and pledged themselves to national power. They made the West their role model, introduced Japan to international trade and sought to internationalize. Their power was not solid: The Boshin War lasted a year and a half against the Tokugawa followers. Even after that, there was insecurity due to lack of money, frequent changes of leadership, the occasional resistance, and the lack of organizational structure.

However, in-depth study of Western political, educational, and economic models began—from the Japanese trip to Europe to the invitation of Western experts to the island nation. Customs revenues were reformed, the semi-feudal provinces were replaced by modern administrative units, land taxes were unified, and education and military conscription became compulsory. Of course, the process was not without its crises, such as the 1877 rebellion over the abolition of the samurai class. However, the last quarter of the nineteenth century saw the spread of railways, a postal network, fireproof brick buildings, modern hospitals, banking, public schools, and an increasing number of literate Japanese becoming accustomed to reading newspapers. Ito Hirobumi, one of the youngest leaders of the Restoration, drafted Asia's first national constitution. The Meiji Restoration set in motion the processes that would bear fruit in Japan's modern dynamism in the early twentieth century [40].

Resistance Long, Determined, Steadfast—and Peaceful

Indian jurist Mahatma Gandhi was one of history's greatest thinkers, a spiritual leader, and a key figure in India's independence movement. After his university years, he traveled to South Africa, where he worked as a legal adviser. It was on a train journey that he was asked to change to third class even though he had a valid first class ticket. This experience of discrimination set him on the path to represent the Indian Hindu minority until his death. Gandhi's philosophy was based on the premise that justice will always prevail, and, therefore, only a struggle waged through passive resistance makes sense. When he returned home from South Africa, India was still part of the British Empire. Gandhi expressed his discontent through hunger strikes, and after the first world war, he encouraged the people of his country to resist and was imprisoned for it. As soon as he was released, he continued his hunger strike—an expression of the importance of repairing and restoring Hindu-Muslim relations. After the second world war, he continued his hunger strikes to draw attention to his movement for Indian independence and an end to religious hostilities.

In 1947, Gandhi's country finally became independent. But the country was divided into India and Pakistan. Gandhi was not in favor of this split and fought for change with further hunger strikes. His whole life and lifestyle was defined by simplicity, and he was puritan in his dress and clothing. Many of his thoughts have become well-known catchphrases, one of the most famous being, "Be the change you wish to see in the world" [41].

Global Strike for the Globe

Gandhi's slogan fits perfectly with the attitude of a young Swedish girl. Born in 2003, the climate change activist Greta Thunberg rose to international prominence in 2019 when she spoke at a UN event (and many other world congresses and summits). In the same year, *Time* magazine named Greta Thunberg its Person of the Year.

The stubborn, unyielding teenager started picketing the Swedish parliament, the Riksdag, on Fridays in 2018, instead of going to school to call on the government to take stronger action on climate change. She founded the Skolstrejk für klimatet movement, mobilizing millions with her personality, communication skills, and persistence.

Climate anxiety is mentioned in connection with her childhood: She is said to have been haunted by the concept of ecological crisis even before she was a teenager. Is it really anxiety or an honest perception of the dramatic nature of the global situation? The question is only partly answered by her famous words urging decision-makers to act:

> But I don't want your hope. I don't want you to be hopeful. I want you to panic. I want you to feel the fear I feel every day. And then I want you to act. I want you to act as you would in a crisis. I want you to act as if our house is on fire. Because it is [42].

These are strong, emotive words. Thunberg can still be seen as a reformer in that she wants to change the existing structure with the aim of improving the political and social conditions of groups of individuals [43].

Reform Versus Revolution

Reforms seek to bring about peaceful change to existing laws, policies, practices, and institutions through constructive debate and confrontation. There will always be reforms because governments and institutions have no choice but to adapt to social change and innovation. Delays in necessary reforms can lead to discontent, tension, and violence.

If governments and institutions are to avoid the need for radical change, they must accept frequent reform as the key to social and political progress. The status quo is easier to gradually repair and change than to overthrow—if politics do not move, the masses will insist on change. In the absence of reform (which, in most cases, can be reversed), irreversible revolutions often undo the status quo and try to create a new and better system. If the change is not drastic, it is easier for people to adapt to it. No one can avoid reform, change, or progress in one form or another. But after a revolution, there is always a legacy of violence, the weight of which must be borne by the various communities in society.

In Sect. 5.1.3, we started to discuss the possible mechanisms of cultural change that might and hopefully should lead to the transition from the throw-away society to the repair society. Can it happen through a peaceful revolution?

Is there such a thing as a peaceful revolution? John F. Kennedy said that "Those who make peaceful revolution impossible will make violent revolution inevitable." Good reform doesn't turn existing systems upside down, nor does it merely tinker at the margins. Instead, it brings about change that then triggers further improvement.

In this book, we are arguing that society as a whole has been sown with the seeds of mental change. New norms of resource management are under development, and there might be mechanisms to implement these norms. So, now we discuss new ways of resource management.

6.4 Toward a Repair Society

6.4.1 How to Repair our Resource Management Strategies

While we are working on this book, we see resources everywhere, from our phones and laptops to our families and colleagues (whom we saw quite infrequently in the last two years) to neighbors and global communities. Resources might exist in mutual relationships. Publishing houses are resources for aspiring authors, for example, but authors are also a resource. If authors would not sign contracts with them, publishing houses could not survive.

There are at least three types of resources. *Private goods* are those whose ownership is restricted to an individual or a group who has already purchased the good.

Your cell phone, a movie ticket you bought (well, before and after a pandemic), and a package of sausage (preferentially *chorizo picante* you bought, maybe too often during the pandemic) are examples. Economists like to use the terms "rivalrous" and "excludable," for situations in which the availability of private goods is finite. The size of a cinema-hall is visibly finite, so access to the resource is excludable based on the purchase of a ticket. If you managed to buy all the tickets, you are the only user. Others, your rivals, can't use it [44].

Public goods are non-rivalrous and non-excludable and available for each member of the society. By society, we mean they are available to the population of a single country, since public goods are determined by national policies and a national budget. Law enforcement and national defense are certainly public goods in each country. Since there is no free lunch, the expenses of the public goods are covered by taxes. National policies decide whether healthcare and public education are public goods. The decisions are far from trivial, and there is a recurring question: Should we or shouldn't we?

Arguments for government spending on public goods state that it is very beneficial for the whole society, since a healthy and well-educated workforce will build a better country. Criticism is based on the argument that taxpayers pay for services they do not necessarily use, and the private sector is more efficient [45].

Common-pool resources are a commodity that, in some sense, are located between a public and private good. A common-pool resource is rivalrous but non-excludable. In principle, it is available to everybody, but its supply is very finite. Fishes in the oceans are typical common-pool resources. Such goods can be overused and over-exploited. While a fisherman—who is not an owner of the resource—knows that more fishing may imply the extinction of fishes, he does not have any incentive to moderate his fishing. If he decides to leave more fish in the ocean today, he will not have more tomorrow, since other fishermen will catch more today. African elephants have also been common-pool resources and are overhunted [46].

The tragedy of elephants is a special case of the *tragedy of the commons*, a concept popularized by the biologist Garrett Hardin (1915–2003) [47]. His paper elaborated on the growing concerns about human overpopulation, but its illustrative example came from English sheep-grazing land. The problem is that each agent would act in his or her own (short-term!) self-interest and consume as much of the commonly accessible, scarce resource as possible. In the longer term, the resource would disappear, leading to tragedy.

Club goods are excludable but non-rivalrous (or at least not until they reach a point where saturation occurs). A private park is a textbook example.

Can We Avoid the Tragedy of Commons?

Three types of solutions to the tragedy of the commons have been suggested [48].

Top-down control of a common-pool resource may prescribe some limits to over-consumption. It is possible to set fish quotas or put some limits on the number of fishing boats or times available to catch fishes. But people are very creative at keeping to the letter of a rule but violating its spirit. For example, if the number of boats is limited, people will build larger boats. So, the regulatory rules will be more and more complicated and inefficient.

Private property rights may create incentives to protect the resource. There is a Hungarian proverb that goes something like "A jointly owned horse has a roughed-up back." People may decide that the common property should be divided for the participants. Grazing commons can be split up and fenced off (even though fences have expenses). Now the individuals have a vested interest in avoiding land degradation. (It is not so easy to build fences for the fishes.) However, the mechanism behind decisions about fair division is far from trivial. Decision making is a political process that results in winners and losers. Human experience suggests that the most disadvantaged people might get very little. In principle, we cannot exclude fair privatization.

An individual transfer quota (ITQ) is a quota imposed on individuals or firms by a governing body that limits the production of a good or service. If the holder of a quota does not use the whole amount of the quota, she has the right to transfer the remaining portion to another participant. Transitioning to ITQs seems to enhance economic performance in the fishing industry but at the cost of coastal communities [49]. While the introduction of ITQs has some definite positive effects, more cooperative solutions look favorable.

Cooperative solutions were suggested by Elinor Ostrom (1933–2012), the first woman to win the Nobel Prize in Economics. She studied community strategies for managing exhaustible resources and found that it is possible to make decisions that ensure long-term protection of resources without government intervention [50]. From her observations, she abstracted a number of rules (sometimes called *Design Principles*), which should and could provide for the healthy functioning of a community [51, 52]. (So, while the rules are normative, there are some—high level—mechanisms to implement them.)

The design principles can be summarized as follows:

1. Define clear group boundaries
2. Match rules governing use of common goods to local needs and conditions
3. Ensure that those affected by the rules can participate in modifying the rules
4. Make sure the rule-making rights of community members are respected by outside authorities
5. Develop a system, carried out by community members, for monitoring members' behavior
6. Use graduated sanctions
7. Provide accessible, low-cost means for dispute resolution
8. Build responsibility for governing the common resource in nested tiers from the lowest level up to the entire interconnected system.

Ostrom's solution emphasizes a bottom-up approach and polycentric governance, rather than the top-down ITQ approach. Apparently, many communities have taken Ostrom's approach and done well for themselves. In contrast to the failure so often observed under state regulation, communal management is highly effective at conserving stocks and maintaining yields. The Maine lobster industry, which produces the about 60% of total U.S. lobster quantities, is a good example of how local communities can play a major role in resource management [53].

Some local communities might care more about fairness than efficiency [49]. It might be better to implement ITQ if there has recently been a large disruption in a region (maybe new fishing industries moving in, maybe local community upheaval). A functional one-size-fits-all approach, backed by government resources, is simpler on average. On the other hand, in a scenario where bottom-up approaches have already been used for decades, it could be irreversibly damaging to enforce a new mode of management for no good reason.

In most of the examples so far, the commodities or resources in question belong to a *local* community—at most to a single country. But there are, as we all know, *global* resources (both public goods and common-pool resources). Now we discuss how to manage these resources at the global level.

6.4.2 Think and Act Glocally!

Global Public Goods

We already know that public goods are *non-rivalrous* and *non-excludable*. If they are available worldwide (while not necessarily to the same extent), they are global public goods. (For further discussion of definitions, see the references [54].) There are also *global commons*, which include the climate, ocean, and air and water quality at large geographical scales. The degradation of these resources happens on an enormous scale, as an indirect consequence of human actions. When traveling faster, disposing of waste, or developing new products for sale, we (and others) do not always see direct, visible consequences. We live in a throw-away society, and the disposal of valueless things has an effects on the quality and quantity of shared resources that are not always immediately visible to us. It is essential to realize that the appropriators of resources are often distant (both spatially and temporally!) from those who lose the most from resource degradation. Large organizations, like corporations and government agencies, and small agents, like individuals and households, might be indirect appropriators. As concerns households in wealthy countries, they are collectively responsible for a significant portion of their nations' emissions.

Paul C. Stern (a former coworker of the previously discussed Ostrom), the president of the Social and Environmental Research Institute, analyzed design principles for global commons and suggested some modifications of the principles Ostrom proposed for more local resources [55]. It was critical that Ostrom and her coworkers (Stern and Thomas Dietz) recognized in a highly cited paper the necessity of adopting the perspective of complex systems research [56]. They offered five principles [55]:

1. Providing good, trustworthy information about resource stocks, flows, and processes and about human-environment interactions affecting the systems;
2. Dealing with conflicts that arise among actors with different perspectives, interests, and fundamental philosophies;

3. Inducing compliance with rules through appropriate combinations of formal and informal mechanisms;
4. Providing physical and technological, as well as institutional infrastructure; and
5. Designing institutions to allow for adaptation.

One of the essential messages is that the appropriate treatment of global public goods and response to common resource problems requires thinking beyond the usual regulatory command and control approaches. Higher-level actors should facilitate the participation of lower-level actors. We discussed in Sect. 6.4.1 the question of how national politics determines the provision of public goods. It is natural to scale up the problem and ask whether institutions can be designed to influence potentially self-interested nation-states to act in service of protecting global public goods. The Intergovernmental Panel on Climate Change (IPCC) is the United Nations body for assessing the science related to climate change [57]. It was founded in 1988 and has 195 member states. Since global warming is a scorching topic (apologies for the pun), design principles for making decisions concerning climate change decisions seem to be interesting for everybody [58].

For the Reader of this book, it might be trivial, but still, we cannot emphasize sufficiently frequently to the public that decision-makers need *science* to understand available resources and patterns of consumption. However, we should acknowledge and accept that there may be a gap between scientific analyses of climate dynamics and the feasibility of needed strategies to cope with anthropogenic climate change. Due to the very nonlinear nature of the physical climate models, any predictability has its internal limits. In addition, the physical climate models are coupled with models for describing human interventions in the climate. We don't have anything better than seeing that "informed decisions require integrating scientific knowledge from quantitative climate models and integrated assessment models with scientific knowledge of coupled human and natural systems, including understandings of political and policy dynamics" [58].

A community of applied scientists who study the principles of managing complex social, ecological, and technological systems recognized the necessity of integrating science, governance, and law in the age when global common resources are changing rapidly [59]. Complex systems science suggests that top-down management of global public goods and common resources is practically impossible. The promise is that self-organized adaptive governance is more appropriate for considering suggestions from local communities.

Glocalization

The expression "globalization" is combination of "globalization" and "local." It was perhaps first used in Japan (Dochakuka), where the relationship between the universal and the particular has a great importance [60]. The term has mostly been used to characterize a business model in which global companies tailor their products and services to take into account the specific demands of the local consumers [61].

The textbook example is how McDonalds conducts business in India. As the cow is considered a holy animal, Indian McDonalds branches remove beef items from the menu and find new ways to satisfy vegetarian consumers. They provide 100% egg-less sandwich sauces, too. McAloo Tikki burgers, consisting of vegetables like a mashed potato patty, are famous.

As another example, some worldwide programs provided the first television coverage of the Olympic Games, and everyone watched the same competitions. It soon was *glocalized*—the Hungarian sport channel, for example, shows more water-polo and canoeing-kayaking than many other national channels.

In early 2022, Oracle opened a data center in South Africa to provide *local* cloud services across Africa for the first time [62]. Global digital infrastructure is now being localized. Global data migrates through the whole world through local data hubs. Africa became Oracle's 37th "cloud region"—an area that allows customers to get faster access from a local data center, in this case, in Johannesburg. As of February 2022, there were 37 cloud regions available and 7 more planned by the end of the year [63].

As I (P) discussed the phenomenon with some of my peers, my mathematician friend (the Readers of the *Ranking* book know him well) gave a beautiful example: When Harry Potter was translated to more than 60 languages, it was glocalized.

According to a well-known phrase, at least in the United States, "all politics is local." But, so far, the glocalization guided by big companies typically implements top-down control.

What about the world? There are now 195 independent sovereign states in the world. Hypothetically, the world could have been governed by a World Government. The concept refers to the idea of all humankind united under a single joint political authority. Proposals for a unified global political power have existed since ancient times but have never materialized. Contemporary political theory and political philosophy offer some normative principles for governing from a global perspective, but there are difficult open questions [64]:

> In 2020 and 2021, as a world divided by deep political, social and economic structural inequalities faces pandemic conditions, economic recession, and environmentally deleterious developments, the questions of whose sense of world community and whose global needs will define the global political agenda and order are more salient than ever.

How Can We Repair Our Damaged Society?

According to the overused quote of Churchill, "Democracy is the worst form of government, except for all the others." Early American democracy was based on a combination of hierarchical and network structures and proved very efficient. While most governments operate in terms of the rule of law, which prescribes that everybody is equally subject to the law, democratic societies are additionally based on the vital principle of free and fair elections. Recent concerns about the crisis facing democracy and democratic elections, magnified by mass protests in response to systemic racism in the United States, can be seen as evidence of an emerging multi-stakeholder fight for a new social order. How can we repair our impaired society?

The goal of repair is not to restore the Golden Age that never was. Sharp debates have already started among those who believe that neither democracy nor capitalism can be thrown out and those who blame the myopic, greedy strategy of the few very rich for eroding democratic freedom.

The way to mend the evil world is to create the right world, as it was written by the famous nineteenth century philosopher Ralph Waldo Emerson (1803–1882) [65]. But we live in the twenty-first century. There is a rapidly formed consensus that the pandemic will change the world forever. The ideas of repair, reform, and even revolution are with us. We don't yet see whether the crisis will strengthen nation-states and reinforce nationalism or make it clear that damages can be repaired through collective global efforts. One might cautiously predict a social order that produces lower profits but provides more stability, resilience, and anti-fragility. Some hope may emerge based on the examples of the power of the human spirit we have recently seen in many countries. Health workers, scientists, politicians, and citizens have shown bravery, demonstrating resilience, effectiveness, and leadership. That reinforces our belief that women and men worldwide will successfully repair the world.

Multilateral Cooperation is Not a Zero-Sum Game, and There Is No Alternative

What is called poly-centric governance (to use Elinor Ostrom's term) appears as multilateral cooperation in the language of international relations. The present global order emerged after World War II through the formation of the United Nations, the International Monetary Fund, World Bank, World Health Organization, and World Trade Organization. Even with all the problems of these organizations, our best chance is to trust them. We might see them as some bottom-up self-organizing mechanisms, but they provide some top-down control, too.

> The concept of the nation-state is in deep crisis. At its core, the nation-state is defined by a geographical border, with governments elected—at least in the context of democracy—to safeguard citizens' interests, improve the quality of available services, manage scarce resources, and promote a gradual rise in living standards. However, as made abundantly clear during the 2008 global financial crisis, economic systems are no longer confined to national borders. They span multiple states in ways that gradually forcing governments to relinquish or share control in a growing number of areas. Indeed, one of the main lessons to emerge from the financial crisis, as noted by former EU Commissioner Peter Mandelson, is that "a global economy needs global economic governance" [66]. The same can be said for the environment and a range of other matters [67].

We know it is a little oversimplified, but it works: Politics is carried out by political institutions and politicians. There is an outstanding question (raised on Quora, which is one of many Internet forums reflecting collective wisdom): "Why do a lot of politicians think that world politics is a zero sum game? Why can't it be a win-win situation for all countries?". One response that we can accept more than half-heartedly was [68]:

One, they are lawyers rather than engineers, and Two, because they are "Social Darwinists" who subscribe to the "Law of the Jungle."

More scientifically, empirical studies and theoretical analysis suggest that politics is considered a zero-sum game [69, 70]. A vote for one candidate means another candidate loses that vote. But positive sum games are better, at least from a global perspective. In a non-zero sum game, there is either a net benefit or a net loss based on the game's outcome.

A book should avoid actualities, but as we write this part at the end of February/early March 2022, Russia's invasion of Ukraine marked the start of a severe geopolitical crisis. Multilateral institutions now have limited power. On the night the UN Security Council met to discuss the growing crisis, Russia announced the start of a *special military operation* and blocked Security Council action on Ukraine. The crisis has spawned no shortage of commentary on the drivers of Russian aggression. As one analysis conjectured:

> Putin's goal is not an occupation of Ukraine but assurances preventing Kyiv's future membership in NATO or the European Union. Unlike China, Russia is in long-term structural decline, unable to wage a long-term overt war against the U.S. or its allies. The United States and Russia will not come to blows over Ukraine because it suits neither power's interest as what happened during Cold War episodes such as the Cuban Missile Crisis. Moscow will reach a mutual understanding with Washington and Brussels, where it withdraws from Ukrainian territory but with agreements allowing Putin to claim an imaginary domestic victory [71].

The international status quo has been considerably perturbed. The Reader knows better than the authors do at this moment whether the perturbation was just a transient deviation from the status quo or a very fundamental change. Abrupt changes in governmental policies (like the outbreak of war) can be explained by negative and positive feedback processes [72]. We have yet to learn how—if at all—the events force the evolution of cooperation among geopolitical stakeholders so as to avoid even bigger future catastrophes. There is no alternative to multilateral cooperation.

6.5 Lessons Learned and Looking Forward

Since the world has never been considered perfect, only in the Golden Age that never was, the concept of "Repair the world" has always been with us. By looking into the past, we see people's wish for rebirth or at least renewal. Grassroots movements can create serious changes, and, as we discussed in Sect. 4.3, the Right to Repair concept now has a fair chance of being accepted by the big tech companies.

At a very global level, we see that it is not possible to deny that existential risk threatens us. We need closed-loop and self-stabilizing technologies, and our best hope is that multilateral negotiations lead to some acceptable solutions to international conflicts.

Past global injustice has left a catastrophic legacy. What should we do? While history cannot be reversed, social tensions should be reduced, since they are drivers of risk and increase the vulnerability of today's society. While we, the authors, don't see a clear mechanism to implement this normative requirement, we trust that the reason and rationality of people will find a solution. Schoolchildren . Gen Z looks different from us. Maybe they will use their knowledge to create a more just world.

The circular economy as a business model of production aims to eliminate waste and pollution. It also tries to regenerate nature in some sense, even though we know that some processes are inherently irreversible. There are now action plans to make sustainable products and ensure less waste. We don't yet see how serious the global effort to implement a circular economy will be. The goal is to provide a resilient system that is good for business, people, and the environment.

In principle, international cooperation regulates the quality and availability of global public goods. It is possible to avoid the tragedy of the commons by combining bottom-up, self-organizing mechanisms with multilateral international agreements. In any case, we need peace.

In early March 2022, Steven Pinker asked whether Russia's war with Ukraine marks the end of the long peace: "No one knows whether it will reverse the Long Peace and send the world back to an age of warring civilizations. Maybe—but maybe not" [73].

References

1. Freeman T (2018) Wisdom to heal the Earth: meditations and teachings of the Lubavitcher Rebbe. Ezra Press and Chabad.org
2. Tikkun Olam. Repairing the World. My Jewish Learning. https://www.myjewishlearning.com/article/tikkun-olam-repairing-the-world/. Accessed 17 Apr 2022
3. Radnoti S. Lecture notes. Eötvös Loránd University (unpublished)
4. Burdick E, Wheeler H (1962) Fail-safe. ECCO
5. Lumet S (director). Fail safe. https://www.tcm.com/tcmdb/title/4556/fail-safe#overview
6. Mian Z (1998, July 11) No time to think. Nuclear Age Peace Foundation. https://www.wagingpeace.org/no-time-to-think/
7. Joint statement of the leaders of the five nuclear-weapon states on preventing nuclear war and avoiding arms races (2022, Jan 3). White House Briefing Room. https://www.whitehouse.gov/briefing-room/statements-releases/2022/01/03/p5-statement-on-preventing-nuclear-war-and-avoiding-arms-races/
8. Gilliland M, Ivanova E (2022) FTP057, 058, 059: Daniel Schmachtenberger—solving the generator functions of existential risks. Future Thinkers Podcast. https://futurethinkers.org/daniel-schmachtenberger-generator-functions/. Accessed 17 Apr 2022
9. Welcome to the Game B Wiki!. Game B Wiki. https://www.gameb.wiki/index.php?title=Main_Page. Accessed 17 Apr 2022
10. Return to begin to repair (2017, Sept 21). Colonialism Reparation. https://www.colonialismreparation.org/en/return-to-begin-to-repair.html
11. Diallo R (2021, July 21) Europe has a long way to go on reparations and making amends for colonialism. Washington Post
12. Getachew A (2020, July 27) Colonialism made the modern World. Let's remake it. The New York Times. https://www.nytimes.com/2020/07/27/opinion/sunday/decolonization-statues.html

13. Menon P (2021, Feb 3) New Zealand plans national syllabus on Maori and UK colonial history. Reuters. https://www.reuters.com/article/us-newzealand-maori/new-zealand-plans-national-syllabus-on-maori-and-uk-colonial-history-idUSKBN2A31G0
14. Vitányi B (2016) A restaurálás rövid története és etikája. Zempléni Múzsa. http://epa.oszk.hu/02900/02940/00064/pdf/EPA02940_zmimuzsa_2016_4_037-039.pdf
15. https://poets.org/poem/archaic-torso-apollos . Accessed 17 April 2022
16. von Foerster H (1960) On self-organizing systems and their environments. In: Cameron S (ed) Yovits MC. Self-organizing systems, Pergamon, pp 31–50
17. Taleb NN (2012) Antifragile: things that gain from disorder. Random House, New York
18. Nygard M (2016, Feb 18) The new normal: from resilient to antifragile. Cognitect. https://www.cognitect.com/blog/2016/2/18/the-new-normal-from-resilient-to-antifragile
19. Niyazov S (2021, Mar 25) The antifragile society. Conjecture Magazine, Medium. https://medium.com/conjecture-magazine/the-antifragile-society-779c3524111f
20. Lacey N, Pickard H (2015) To blame or to forgive? Reconciling punishment and forgiveness in criminal justice. Oxford J Legal Stud 35(4):665–696. https://doi.org/10.1093/ojls/gqv012
21. Restorative justice network. http://restorativejustice.org/. Accessed 17 Apr 2022
22. Naar D (2021) What are the pros and cons of restorative justice? Reference. https://www.reference.com/world-view/pros-cons-restorative-justice-a722d3404aa5cb87. Last updated 11 May 2021
23. Korhonen J, Honkasalo A, Seppälä J (2018) Circular economy: the concept and its limitations. Ecol Econ 143:37–46. https://doi.org/10.1016/j.ecolecon.2017.06.041
24. Nelson A, Coffey B (2019, Nov 4) What is 'ecological economics' and why do we need to talk about it? The Conversation. https://theconversation.com/what-is-ecological-economics-and-why-do-we-need-to-talk-about-it-123915
25. Wreg R (2022) 12 Examples of circular economy solutions that give us some hope! Innovate Eco. https://innovate-eco.com/circular-economy-solutions/. Accessed 17 Apr 2022
26. https://www.youtube.com/watch?v=UaHTnCeVZl0
27. Simke A (2019, Dec 1) Black soldier flies are the new superstars of sustainable aquaculture. Forbes. https://www.forbes.com/sites/ariellasimke/2019/12/01/black-soldier-flies-are-the-new-superstars-of-sustainable-aquaculture/
28. Stumpf C (2020, June 4) Sustainable solutions for plastics: the future role of lignins. Chemistry World. https://www.chemistryworld.com/future-of-plastics/sustainable-solutions-for-plastics-the-future-role-of-lignins/4011826.article
29. It's the end of the road for waste plastics. MacRebur: The Plastic Road Company. https://macrebur.com/us. Accessed 17 Apr 2022
30. https://www.youtube.com/watch?v=yfGZDPJNAaM
31. Mak SL, Wu TMY, Tang FWF, Li, JCH, Lai CW (2021) A review on utilization of plastic wastes in making construction bricks. Presented at IOP Conference series: Earth and environmental science, vol 706. https://iopscience.iop.org/article/10.1088/1757-899X/691/1/012083/pdf
32. Garcia C (2021, Feb 4) Kenyan woman finds a way to recycle plastic waste into bricks that are stronger than concrete. Yahoo! News. https://news.yahoo.com/kenyan-woman-finds-way-recycle-053411779.html?_sp=bc83c517-7fa8-4722-abeb-e2cb768122fa.1650254584567
33. Morris DZ (2016, Mar 13) Today's cars are parked 95% of the time. Fortune. https://fortune.com/2016/03/13/cars-parked-95-percent-of-time/
34. Sposato P, Preka R, Cappellaro F, Cutala L (2017) Sharing economy and circular economy. How technology and collaborative consumption innovations boost closing the loop strategies. Environ Eng Manag J 16(8):1797–1806. http://www.eemj.icpm.tuiasi.ro/pdfs/vol16/no8/17_61_Sposato_17.pdf
35. Reform. Definition from Oxford languages, accessed via Google. https://www.google.com/search?client=firefox-b-d&q=reform+meaning. Accessed 2022, Apr 2017
36. Gombrich E (1935) Weltgeschichte von der Urzeit bis zur Gegenwart. Steyrermühl-Verlag, Wien-Leipzig
37. Julian calendar. Britannica. https://www.britannica.com/science/Julian-calendar. Last updated 1 July 2019

38. Teall J, Nicol DM (2021) The reforms of Diocletian and Constantine. Britannica. https://www.britannica.com/place/Byzantine-Empire/The-reforms-of-Diocletian-and-Constantine. Last updated 21 Sept 2021
39. Reformation. https://www.luther2017.de/reformation/
40. Huffman J (2022) The Meiji Restoration era, 1868–1889. About Japan. http://aboutjapan.japansociety.org/content.cfm/the_meiji_restoration_era_1868-1889#sthash.viQmeC3O.dpbs. Accessed 17 Apr 2022
41. Mahatma Gandhi. Wikipedia. https://en.wikipedia.org/wiki/Mahatma_Gandhi. Last updated 17 Apr 2022
42. Workman J (2019, Jan 25) Our house is on fire. 16 year-old Greta Thunberg wants action. World Economic Forum. https://www.weforum.org/agenda/2019/01/our-house-is-on-fire-16-year-old-greta-thunberg-speaks-truth-to-power/
43. Kraemer D (2021, Nov 5) Greta Thunberg: who is the climate campaigner and what are her aims? BBC News. https://www.bbc.com/news/world-europe-49918719
44. Chen J (2021) Private good. Investopedia. https://www.investopedia.com/terms/p/private-good.asp. Last updated 5 Jan 2021
45. Fernando J (2022) Public good. Investopedia. https://www.investopedia.com/terms/p/public-good.asp. Last updated 20 Mar 2022
46. Jung BD (2017) The tragedy of the Elephants. Wisconsin Law Rev 2017(4). https://repository.law.wisc.edu/s/uwlaw/item/299879
47. Hardin G (1968) The tragedy of the commons. Science 162(3859):1243–1248. https://doi.org/10.1126/science.162.3859.1243
48. Tragedy of the commons. Investopedia. https://www.investopedia.com/terms/t/tragedy-of-the-commons.asp. Last updated 12 Mar 2022
49. Eythorsson E (1996) Coastal communities and ITQ management. The case of Icelandic fisheries. Sociol Ruralis 36(2):212–223. https://doi.org/10.1111/j.1467-9523.1996.tb00017.x
50. Ostrom E (1990) Governing the commons: the evolution of institutions for collective action. Cambridge University Press, Cambridge
51. Becker CU (2019) Ethical principles for design. Presented at the International association of societies of design research conference 2019. https://iasdr2019.org/uploads/files/Proceedings/vo-f-1307-Bec-C.pdf
52. White N (2014, Nov 7) Elinor Ostrom's 8 principles for managing a commmons—words to live by. Full Circle Associates. https://fullcirc.com/2014/11/07/elinor-ostroms-8-principles-for-managing-a-commmons-words-to-live-by/
53. Aligica PD, Sterpan I (2017) Governing the fisheries: insights from Elinor Ostrom's work. In: Wellings R (ed) Sea change. How markets and property rights could transform the fishing industry, Institute of Economic Affairs, pp 95–116
54. Ress MA (2013, Aug 23) Global public goods, transnational public goods: some definitions. Knowledge Ecology International. https://www.keionline.org/book/globalpublicgoodstransnationalpublicgoodssomedefinitions
55. Stern PC (2011) Design principles for global commons: natural resources and emerging technologies. Int J Commons 5(2):213–232. https://doi.org/10.18352/ijc.305
56. Dietz T, Ostrom E, Stern PC (2003) The struggle to govern the commons. Science 302(5652):1907–1912. https://doi.org/10.1126/science.1091015
57. The intergovernmental panel on climate change. https://www.ipcc.ch/. Accessed 17 Apr 2022
58. Stern PC, Wolske KS, Dietz T (2021) Design principles for climate change decisions. Curr Opin Environ Sustain 52:9–18. https://doi.org/10.1016/j.cosust.2021.05.002
59. Cosens B, Ruhl JB, Soininen N, Gunderson L, Belinskij A, Blenckner T, Camacho AE, Chaffin BC, Craig RK, Doremus H, Glicksman R, Heiskanen AS, Larson R, Similä J (2021) Governing complexity: integrating science, governance, and law to manage accelerating change in the globalized commons. Proc Natl Acad Sci U S 118(36):e2102798118. https://doi.org/10.1073/pnas.2102798118
60. Robertson R (1995) Glocalization: time-space and homogeneity-heterogeneity. In: Featherstone M, Lash S, Robertson R (eds) Global modernities. Sage Publications, Thousand Oaks, pp 25–44

61. Roudometof V (2015) The glocal and global studies. Globalizations 12(5):774–787. https://doi.org/10.1080/14747731.2015.1016293
62. Mukherjee S, Mukherjee P (2022, Jan 19) Oracle opens data centre to provide cloud services across Africa. Reuters. https://www.reuters.com/technology/oracle-opens-data-centre-provide-cloud-services-across-africa-2022-01-19/
63. Oracle cloud regions. Oracle. https://www.oracle.com/cloud/architecture-and-regions/. Accessed 17 Apr 2022
64. Lu C (2021) World government. In: Zalta EN (ed) The Stanford encyclopedia of philosophy, spring 2021 edn. https://plato.stanford.edu/entries/world-government/. Last updated 5 Jan 2021
65. Emerson RW (1860) The conduct of life. https://www.gutenberg.org/ebooks/39827
66. Mandelson P (2008, Oct 3) In defence of globalization. The Guardian
67. Lopez-Claros A, Dahl A, Groff M (2020) The challenges of the 21st century. Global governance and the emergence of global institutions for the 21st century. Cambridge University Press, Cambridge, pp 3–29
68. Chu B (2020) Why do a lot of politicians think the world politics is a zero sum game? Why can't it be a win-win situation for all countries? Quora. https://www.quora.com/Why-do-a-lot-of-politicians-think-the-world-politics-is-a-zero-sum-game-Why-can-t-it-be-a-win-win-situation-for-all-countries
69. Davidai S, Ongis M (2019) The politics of zero-sum thinking: the relationship between political ideology and the belief that life is a zero-sum game. Sci Adv 5(12). https://doi.org/10.1126/sciadv.aay3761
70. Brunnermeier M (2021) The resilient society. Endeavor Literary Press
71. Brown K (2022, Feb 25) The thucydides trap comes for Russia and America 1945. https://www.19fortyfive.com/2022/02/the-thucydides-trap-comes-for-russia-and-america/
72. Bumgartner FK, Jones FD (eds) (2002) Policy dynamics. University of Chicago Press, Chicago
73. Pinker S (2022, Mar 2) Is Russia's war with Ukraine the end of the long peace? The Boston Globe. https://www.bostonglobe.com/2022/03/02/opinion/is-russias-war-with-ukraine-end-long-peace/

Chapter 7
Epilogue: Toward a Repair Society

Abstract In the closing chapter, we summarize our statements about the scope and limits of reparability. By repairing, we extend the life of things, save money, and reduce our ecological footprint. People are often faced with a dilemma regarding when to attempt to repair and when to let things go. In any case, we discussed repair mechanisms that occur at very different levels of social organization, from individual objects to individual people, from our relationships to our society as a whole. We should throw out the Throw-away society and replace it with the Repair society.

We Must Realize the Repair Society

"What cannot be repaired is not to be regretted!" When we started to think about writing this book together, we soon agreed that the one slogan of the book should come from Samuel Johnson, the celebrated eighteenth century English author. But we need other slogans to motivate us to take action. Emerson's "The way to mend the bad world is to create the right world" seems to be still relevant as well.

Our goal with *Repair* was to propagate a new way of thinking about resource management. By resources, we meant not only gadgets, like cell phones and cars, but also human resources, like our family members, friends, and the small and large communities to which we belong. By repairing, we extend the life of things, save money, and reduce our ecological footprint. People are often faced with a dilemma regarding when to attempt to repair and when to let things go. In any case, we discussed repair mechanisms that occur at very different levels of social organization, from individual objects to individual people, from our personal relationships to our society as a whole. We already know the trichotomy we face: we may try to restore the original state; we may replace the old with a new; we should try to make things better than they were.

As one of our heroes, Thomas Schelling, showed in his bestseller *Micromotives and Macrobehavior*, small and seemingly senseless decisions and actions by individuals often lead to unintended dramatic consequences for smaller or larger communities [1]. We attempted to demonstrate how general concepts (e.g., stability and resilience) work at the different levels of analysis we have considered.

It would be difficult to deny that there was a rapid transition (of course, on a historical scale) from general poverty to a throw-away society. We cannot deny that inequality increases while the middle class likes consumption. A significant proportion of the human population can afford (in a strictly monetary sense) food and textile waste. But it turns out that unrestricted waste is irresponsible and has dramatic consequences for our physical and social environments. The continuation of present tendencies will lead us to ecological, economic, political, and social crises. Society should make fast and responsible decisions locally and globally in the transition to what we may call a repair society.

Nostalgia-Driven Repair

The book started with some of our own nostalgic personal stories, which reflect our early encounters with the concept of repair. Developments in science and technology were the driving force of the transition from postwar poverty to the throw-away society. However, after several decades it became visible that the trajectory was not sustainable.

The nostalgic part of us likes to think about the "Age Before Things Went Wrong." Nostalgia was initially considered a mental illness. The attitude changed from the middle of the nineteenth century, when it was transformed from a dangerous but curable illness into a state of the non-pathological mind. Research shows that twenty minutes of nostalgia per day can make us feel happier and more cheerful [2]. (But remember the Proust quote from Sect. 2.6: "Remembrance of things past is not necessarily the remembrance of things as they were." In critical situations, we like to believe that it is possible to return to previous, happier times. Therefore, we leave to the Reader to decide how her memories about the past influence her present decisions about saving and repairing.

Les Fleurs du Mal

Next, we discussed the different mechanisms by which things go wrong and fall apart in nature and society—the *Flowers of Evil*, as labeled by Charles Baudelaire (1821–1867).

Complex systems theory offers a means to help us understand the mechanisms behind natural and social crises (such as earthquakes, tsunamis, and financial collapse) and the occurrence of extreme events. Due to slowly accumulating, seemingly invisible sub-terrestrial events, many such phenomena emerge. At the level of social relationships, writers and social psychologists offer explanations for the different mechanisms underpinning the decay of friendships, which are also a kind of disaster, at least for the people in question.

The rules related to disasters are built-in features of nature and society. They are connected to the conflict between constancy and change, demonstrated in several disciplines from physics to economics to social psychology. We have the first and second laws of thermodynamics, which express these principles. Specifically, the first specifies the conservation of energy, and the second the rules of irreversibility. In other words, the amount of energy in isolated systems is constant, but a system always tends toward irreversible change. Economic growth does not appear to be reconcilable with sustainability. In our current society, we have seen unsustainable levels of resource consumption—especially of resources that cannot be renewed—in the name of economic growth.

Stability, Homeostasis, Resilience, and Their Enemies

Repair develops our imagination, critical thinking, and creativity. The simplest repair mechanisms find pathways back to "normal." When a system deviates just slightly from its normal state, we can expect that this state can be reached again. Scholars in the context of physico-chemistry have elaborated concepts related to stability, and we studied the analogs of this concept in the social sciences. This does not mean we are physicalists in the negative sense of the word. Complex adaptive systems might exhibit the plasticity-stability (or plasticity-rigidity) dichotomy, which is related to the concept of system stability. Negative feedback loops are associated with stability, since they tend to bring a process to equilibrium (while positive feedback loops tend to accelerate a process and move a system away from equilibrium). The concept of homeostasis has proven a helpful analogy in the social sciences.

A person can maintain a stable psychological condition under conflicting stresses and motivations. A society homeostatically maintains its stability (well, for a while!), despite competing cultural, economic, and political factors. One of the main ingredients of classical economic theory is the law of supply and demand, whereby the interactions of consumers and the resulting price changes have a stabilizing effect.

Resilience has become a buzzword that expresses that systems at very different levels of biological and social organization can react to disturbances and maintain their basic function. Individuals, buildings, communities, and local and global communities might preserve their functions after shock. How? To be more precise: We investigated the question of how to design resilient systems.

We are finishing writing the book in March 2022, when the Russian invasion of Ukraine created a vast disturbance in the world order, so what we have is the *new abnormal*. While we wrote this book intending to radiate some optimism, our hearts are nearly broken. A low probability and high impact shock is unfolding before our eyes. The world will need more repair than ever before.

We are not very happy that our theoretical analysis about uncompensated positive feedback in Sect. 3.3.2 and bistable systems in Sect. 5.1.1 can be applied to the present, real-world context of war. However, they have explanatory power. Russian military action is answered by the "West" through economic punishment up to unprecedented levels that may implement mutual positive feedback, as we described in Sect. 4.1.1. On March 3, 2022, Christopher Chivvis, the director of the American Statecraft Program at the Carnegie Endowment for International Peace, in his article "How Does This End?" predicted two plausible scenarios [3]: "one, continued escalation, potentially across the nuclear threshold; the other, a bitter peace imposed on a defeated Ukraine that will be extremely hard for the United States and many European allies to swallow." To formulate it in more journalistic language: The start of World War III or a new Cold War [4].

While in Chap. 4 we analyzed the pathways *back* to the "normal" that was, Chap. 5 investigates situations when the perturbation exceeds the region of stability, and the systems move towards uncharted territories, called the new (ab)normal.

Repair Society: Good Morals and Technological Development

We need good morals to help decide when something is repairable and if it should be repaired. We also need new technology to help us achieve our goals. Technology is neutral. It can improve human physical, mental, and moral well-being. It can help people become healthier and more educated and make better ethical decisions. However, it can also do the opposite: It can make us sicker, less educated, less loving of others, and worse at making moral decisions.

Principles of Repair:

1. A. Repair! B. Replace! C. Create something new!
2. By repairing, we extend the life of things, save money, and reduce our ecological footprint.
3. Repairing develops our imagination, critical thinking and creativity.
4. Through repair we discover: objects, materials, processes and more.
5. What we repair, we enrich. A repaired object becomes unique, and this uniqueness outlives fashion.
6. Repair makes us independent: We are less vulnerable to manufacturers.
7. What can be repaired must be repaired.

Theory suggests that complex systems are more vulnerable than simple ones. Failures may occur due to a high degree of connectivity. We also know that insufficient negative feedback may lead to natural and social disasters. The good news is that we are already in a transition period where a significant minority is beginning to recognize that both human and physical resources are worth managing.

At the individual and community level, we need to figure out how to fight against the constant need to replace new things. After a certain age, we also realize that we have to look after our friendships. On the one hand, we need more durable products. But, on the other hand, we need to accept that broken items can and should be repaired.

As concerns the whole human society: It is impossible to continue this path without repeating the tragedy of the civilization of the Easter Islands on a much larger scale. We should throw out the Throw-away society and replace with the Repair society.

References

1. Schelling TC (2006) Micromotives and macrobehavior. W. W. Norton & Company. (Original work published 1978)
2. Krakovsky M (2006) The art of remembrance. Psychol Today. https://www.psychologytoday.com/us/articles/200605/the-art-remembrance
3. Chivvis CS (2022) How does this end? Carnegie endowment for international peace. https://carnegieendowment.org/2022/03/03/how-does-this-end-pub-86570
4. Hauck G (2022) A new cold war, or the start of world war III? How historians see the invasion of Ukraine. *USA Today*. https://www.usatoday.com/story/news/nation/2022/02/24/cold-war-wwiii-russia-ukraine/6923412001/?gnt-cfr=1

Index

A
Ancient book repairing in China, 119
Anti-fragility, 132, 146
Arrow of time
  - biological, 44
  - cosmological, 54
  - historical, 54
  - thermodynamic, 53, 107
Artificial Intelligence (AI), 9, 58, 77, 127

B
Biodegradable, 17, 134
Bounded confidence, 112
Burnout, 51, 56, 86

C
Catastrophe theory, 104, 105
Cellular automata, 108
Circular economy, 21, 22, 133, 135, 148
Climate crisis, 43–45, 79, 80
Cognitive diversity, 29
Collapsology, 58
Collective behavior, 111
Collective phenomena, 63
Common-pool resources, 141, 143
Complex systems theory, 59, 79, 91, 115, 154
Concorde effect, 117
Consensus, 27, 43, 112, 146
Conservation of energy, 52, 154
Cooperative solutions, 142
COVID-19, 13, 20, 56, 58, 71, 76, 83, 91, 92, 108, 116, 121
Creative destruction, 8, 90, 113
Cultural changes, 114, 115, 135, 140

Cybernetics, 77, 98, 131

D
Decision making, 27, 87, 116, 142
Degrowth movement, 62
Democracy, 37, 39, 43, 145, 146
Destruction of groups, 71
Disasters, 6, 7, 18, 29, 39, 44, 51, 56, 69, 72, 75, 76, 81, 84, 85, 87–90, 98, 110, 122, 154, 156
Disinhibition-induced epilepsy, 65
Diversity, 28, 29, 108, 112
Doughnot economic model, 62
Drought, 81, 83
Dunbar number, 110
Dynamical systems, 26, 107
Dynamic models of segregation, 108

E
Echo chambers, 27
Economic growth, 20, 21, 43, 61, 79, 104, 154, 155
Entropy, 52–54
Epilepsy, 63, 64
Epileptogenesis, 64
Ethics of restoration, 130
Evolutionary psychology, 26
E-waste, 11, 12, 94, 98
Existential risk, 72, 125, 127, 147
Exponential growth, 9, 61, 127
Extreme events, 59, 60, 62, 65, 72, 154

F
Fast fashion, 7, 18, 19, 21, 134

© The Editor(s) (if applicable) and The Author(s), under exclusive license to Springer
Nature Switzerland AG 2022
P. Èrdi and Z. Szvetelszky, *Repair*,
https://doi.org/10.1007/978-3-030-98908-8

Feedback
- loops, 8, 60, 78, 79
- negative, 56, 60–63, 75, 78–80, 98, 127, 155, 156
- positive, 27, 56, 60, 61, 63, 64, 66, 71, 77, 79, 80, 110, 111, 147, 155
- uncompensated, 60, 63, 72, 78, 155
Feeding America, 14
Financial crisis, 6, 63, 146
Finite time singularity, 61
First and second laws of thermodynamics, 52, 72, 154
Forest fire, 7, 115
Fragmentation, 112
Fragments, 130, 131
Friendship, 6, 23–25, 29, 66, 70, 72, 116, 117, 154, 156

G
Garden of Eden, 35, 40, 41, 47
Global inequality, 9
Global injustice, 125, 128, 130, 147
Global public goods, 143, 144, 148
Glocalization, 144, 145
Goal-oriented behavior, 77
Golden age, 3, 30, 35–47, 63, 87, 145, 147
Golden repair, 119
Good morals, 132, 133, 156
Groupthink, 29
Growth
- exponential, 9, 61
- linear, 61
- super-exponential, 9, 61, 62, 127
- unbounded, 60, 61, 107

H
Heat waves, 81
Hedgehogs and foxes, 28
History of reforms, 136
Homeostasis, 70, 75, 77, 79, 98, 155
Hurricane Katrina, 89, 90, 113

I
996.ICU, 56
Imperfect food, 15
Industrial revolution, 7, 8, 14, 16, 19, 58
Inequality of wealth, 79
Information sharing, 29
Iron age, 40, 44
Irreversibility, 51–54, 154

K
Kintsugi, 119–121

L
Le Chatelier's principle, 75–77, 83, 98
Les Fleur du Mal, 154
Limits to growth, 62
Linear economy, 21, 22
Little Ice Age, 45

M
Mathematical models, 9, 105
Meiji restoration, 136, 138, 139
Microplastics, 17
Misfit market, 15
Multilateral cooperation, 146, 147

N
Neoclassical economic theory, 61
Netflix, 86, 113
New normal, 80, 103–105, 107, 112–114, 119, 121, 122, 132
Nostalgia, 44, 46, 47, 154

P
Pandemic, 2, 6, 13, 14, 18, 54, 56, 58, 61, 71, 76, 77, 81, 85, 89, 91, 92, 108, 110, 113, 121, 131, 141, 145, 146
Paradise Lost, 40
Paradise Regained, 40
Planned obsolescence, 7, 10–12, 23, 97, 115

Plastic, 12, 17, 20, 134, 135
Polarization, 27, 51, 71, 112
Post-colonialism, 128
Private goods, 140, 141
Private property rights, 142
Problem solving, 27, 28, 58
Public goods, 141, 143, 144

R
Reform versus revolution, 140
Repair, 1, 3, 5–7, 11, 12, 14, 22, 23, 25, 29, 30, 35, 36, 39, 47, 51, 53, 57, 70, 73, 75–77, 81, 83, 86, 88, 90, 93–99, 103, 114–120, 122, 125, 126, 128, 130, 132, 133, 135, 136, 140, 145–147, 153–156

Repair manifesto, 96
Repair society, 6, 99, 115, 140, 153, 156

Replace of repair dichotomy, 22
Residential segregation, 109
Resilience, 75, 80–85, 89, 90, 92, 98, 146,
        153, 155
Resilient society, 85
Resource management, 1, 22, 140, 143, 153

Resource management strategies, 140
Resources
 - physical, 24, 156
 - social, 6, 7, 23, 24
Restorative justice, 132, 133
Restoring buildings, 86
Right to repair, 6, 23, 93–97, 99, 118, 147

S
Self-organization
 - social, 105, 110
Sharing economy, 18, 19, 135
Social group, 26, 28, 35, 51, 110
Social relationships, 23, 30, 110, 154
Societal collapse, 57, 58, 72
Sociobiology, 25, 26
Spatial patterns, 108, 109
Spontaneous glass breakage, 54, 55
Stability, 30, 35, 37, 55, 66, 75–77, 79, 80,
        98, 107, 146, 153, 155, 156

Stability-plasticity dilemma, 76, 77
Stable relationships, 66, 72
Synchronization, 26, 27, 64
Synthetic fibers, 16, 17

T
Technological singularity, 127
Temporal patterns, 107
Thresholds, 64, 83, 103, 114, 155
Throw-away - repair dilemma, 93
Throw-away society, 1, 6, 7, 11, 20, 21, 30,
        99, 103, 114, 116, 122, 134, 135, 140,
        143, 153, 154, 156
Tikkun Olam, 125, 126
Tipping points, 79, 80, 103, 114, 115
Tragedy of commons, 125, 141
Transformative AI, 58
Tulip Bulb Mania, 63, 64
Two cultures, 52

V
Value engineering, 12

W
Wear and tear, 51, 54, 55, 72, 118, 120

Printed in the United States
by Baker & Taylor Publisher Services